SUPERSPECIFIC GROUPS OF CRETACEOUS AND CENOZOIC TELLINIDAE

The Geological Society of America, Inc.
Memoir 119

Taxonomic Revision of the Superspecific Groups of the Cretaceous and Cenozoic Tellinidae

FREYDOUN AFSHAR

Ankara, Turkey

1969

PUBLISHED BY

THE GEOLOGICAL SOCIETY OF AMERICA, INC.
Colorado Building, P.O. Box 1719
Boulder, Colorado 80302

Printed in the United States of America

The Memoir Series
of
The Geological Society of America
is made possible
through the generous contributions of
Richard Alexander Fullerton Penrose, Jr.,
and is partially supported by
the National Science Foundation.

Contents

PLATE

Preface

When Dr. Afshar left Washington in 1952 to return to Iran, he left his manuscript in my hands with the request that I see it through the various necessary steps leading to its appearance in print. It is unfortunate that for a number of reasons this manuscript has remained virtually untouched, for the considerable labor that has gone into this work should have been available to workers for some time.

In preparation for its publication at this time I have come across a number of superspecific groups that have been proposed since Dr. Afshar finished his manuscript, and these I have added as a postscript. In each case I have copied the original diagnosis, including the description of the type-species, wherever it was given by the author, and whatever remarks were added. To this I have added in most cases my own comments regarding the allocation of the group, and wherever possible its position in Dr. Afshar's system. In all cases I have illustrated the type-species on four supplementary plates either by photographs of the specimens in the collections of the U. S. National Museum or by copies of published figures. These new names have also been incorporated in the keys prepared by the author.

HARALD A. REHDER

Acknowledgments

This study was made under the direction of Professor H. E. Vokes of The Johns Hopkins University, who proposed the problem and to whom the writer is indebted for constant advice, generous help and constructive criticism, and for making available to the writer his personal library. The writer is deeply grateful to Dr. Julia A. Gardner of the United States Geological Survey for helpful advice and criticism and for her courtesy in permitting the use of some rare publications on the subject from her personal collection.

The writer wishes to express his profound gratitude and indebtedness to Professor J. T. Singewald, Jr., the Chairman of the Department of Geology, for offering the opportunity to pursue this study in his department.

To Dr. H. A. Rehder, Curator of the Division of Mollusks, United States National Museum, is due my most sincere appreciation for his personal guidance and advice on the problem and for offering the facilities of the Museum—without them this study would not have been possible.

The writer is also grateful to Dr. J. P. E. Morrison of the U. S. National Museum, Dr. R. Tucker Abbott of the Academy of Natural Sciences of Philadelphia, and to Dr. F. S. MacNeil, formerly of the U. S. Geological Survey, for the generous help and advice which they offered during the writer's study at the Museum. He desires to express his gratitude also to the late Dr. J. B. Reeside, Jr., of the U. S. Geological Survey for permitting him to study the Cretaceous fossils that were included in this work, to the late Dr. L. W. Stephenson of the U. S. Geological Survey for permitting the writer to copy some of the illustrations from his publications on Cretaceous faunas, and to Dr. W. P. Woodring of the U. S. Geological Survey for making his personal library available to the writer. The writer is grateful to Dr. W. J. Clench of Harvard University, the late Dr. H. A. Pilsbry, Mr. A. A. Olsson of Coral Gables, Florida, and to Miss J. Allan of Australian Museum, Sydney, Australia, for lending certain specimens for study.

Introduction

The pelecypod family, Tellinidae, constitutes one of the most distinctive groups of the shallow water marine molluscan faunas. It is represented in all of the oceans and seas from the polar to the tropical regions and within these waters is found from "the level of high water of neap tides" (Yonge, 1949, p. 234) to depths in excess of 600 fathoms (Dall, 1889, p. 60). From within this wide range of geographic and bathymetric distribution, a large number of species and varieties have been described. An investigation of paleontologic literature reveals an apparently equally cosmopolitan distribution and abundance of species during Cretaceous and Cenozoic times.

A number of supraspecific units have been recognized and named from within this complex of species, but, lacking a satisfactory analysis and synthesis of these names of higher category, there has long been a tendency to use a few generic names in a wholly uncritical and unwarranted broad sense that has ignored significant morphologic characters and obscured phylogenetic relationships. Thus almost all elongate forms are referred to the genus *Tellina* Linnaeus if they possess lateral teeth, and to *Macoma* Leach if laterals are absent; the ovate to rounded types have been referred to *Arcopagia* (Leach) Brown, and those with a strong median plication to *Apolymetis* Salisbury (*Metis* Adams, *et auct.*).

The present study originally had as its objectives: (a) the redescription from specimens of the type species of those valid supraspecific units that are properly to be referred to the family Tellinidae; (b) the erection of a system of classification for these units that is based upon significant morphologic characters of the shell and which will, therefore, be usable by the paleontologist; (c) the assignment to those supraspecific units so recognized and classified of the large number of species and varieties that have been described from the Cretaceous and Cenozoic deposits of North America.

It was, unfortunately, not possible to accomplish all of these objectives at this time. Despite the large collections that were made available through the courtesy of the United States National Museum and other depositories, specimens of the type species of a few of the supraspecific units could not be obtained. In these cases the accompanying descriptions are based upon the original illustrations of the species together with such available subsequent illustrations as were considered to represent with certainty the species in question.

1

Furthermore, the specimens of a great number of the species were not available; even if all had been available, due to lack of time, I would not have been able to examine and study every species in this family and place them all in the correct groups. For this reason it was deemed inadvisable to attempt the last of the three objectives outlined above.

Abstract

Earlier attempts at a classification of the family Tellinidae are reviewed, and a classification of the family based on shell characters is proposed. These essential characters are the nature of the hinge teeth and the nature of the pallial sinus; characters of secondary importance are the shape of the shell, sculpture, relative position of the umbones and size of the shell.

The family is divided into fifteen genera, with ten of these being further divided into subgenera; all in all, 101 superspecific taxa are proposed, of which ten are new. Keys to the genera and almost all of the subgenera are provided.

A diagnosis of each genus and subgenus is given, together with bibliographic references, citation of the type-species, and geologic range of the taxon, and a description and figure of each type species.

In a supplement, the same information is given for twelve superspecific taxa proposed since the completion of the work by the author in 1952. Comments are given as to the probable location of these groups within the author's classification.

The Generic Names in the Tellinidae

The generic name *Tellina* is one of the oldest in conchological literature and was used by many authors before Linnaeus, notably by Rumphius (1711; 1741), Gualtieri (1742), d'Argenville (1742; 1755), Klein (1753), and Adanson (1757). Not all of these earlier usages agree with that accepted by Linnaeus; Adanson, for example, in his "Histoire Naturelle du Senegal: Coquillages, and so on." used the name for a group of related species that corresponds almost exactly with that to which Linnaeus in 1758 applied the generic name *Donax*.

In the tenth edition of his "Systema Naturae" Linnaeus referred 25 recent species to *Tellina*. In 1866 Reeve, in the "Conchologia Iconica" has 345 species; and at the present time, more than 500 Recent species and varieties are referable to the family Tellinidae, which is approximately synonymous with the genus *Tellina* as originally used by Linnaeus.

Lamarck in 1799 in his "Prodrome d'une Nouvelle Classification des Coquilles" cited *Tellina virgata*, the third species in the Linnaean list, as an "example" to illustrate the genus. This was not a designation of a type for the genus (*see* Opinion 79, International Rules of Zoological Nomenclature, which, though specifically applied to Lamarck's "Systeme des Animaux sans Vertebres" of 1801 seems equally applicable to his earlier usage of "examples") but it was so accepted by a number of authors, especially by Dall (1900b, p. 289) who has made one of the more careful attempts to classify the members of the Tellinidae.

The first apparently valid designation appears to be that of Schmidt, 1818, who selected *Tellina radiata*, the tenth species on the Linnaean list, as the type of the genus *Tellina*. Children, in 1823, chose the same species, as did Gray, 1847, and Stoliczka, 1870. The latter author at the same time designated *Tellina virgata* Linnaeus as the type of *Tellinella* "Gray" Mörch. As noted by Gardner (1928, p. 190), the name *Tellinella* was credited by Mörch to "Gray, 1852." No mention of the name can be found in any of Gray's publications and must, therefore, have come from a manuscript or label. It cannot be attributed to Gray.

The 25 species referred to the genus *Tellina* by Linnaeus constituted a rather heterogeneous assemblage of forms, as may be seen by an examination of the accompanying tabular listing of these species together with their present taxonomic assignment. The majority of the species are still to be found within the confines of the family Tellinidae, although only one is now referred to *Tellina* sensu stricto. Some, however, are far removed.

5

PRESENT SYSTEMATIC ASSIGNMENT OF LINNAEAN SPECIES

LINNAEAN SPECIES	PRESENT SYSTEMATIC ASSIGNMENT	
Tellina gargadia	Quadrans (Quadrans) gargadia—type	Tellinidae
Tellina lingua-felis	Arcopagia (Smitharcopagia) linguafelis—type	Tellinidae
Tellina virgata	Tellina (Tellinella) virgata—type	
Tellina gari	Gari (Gari) gari—type	Gariidae
Tellina fragilis	Gastrana (Gastrana) fragilis—type	Tellinidae
Tellina albida	Indeterminable	
Tellina foliacea	Phylloda foliacea—type	Tellinidae
Tellina planata	Tellina (Peronaea) planata—type	Tellinidae
Tellina laevigata	Tellina (Scrobiculina) laevigata	Tellinidae
Tellina radiata	Tellina (Tellina) radiata type	Tellinidae
Tellina rostrata	Tellina (Dallitellina) rostrata—type	Tellinidae
Tellina trifasciata	?Donax trifasciata	?Donacidae
Tellina incarnata	Angulus incarnata	Tellinidae
Tellina donacina	Tellina (Moerella) donacina—type	Tellinidae
Tellina balaustina	Arcopagia (Arcopella) balaustina—type	Tellinidae
Tellina remies	Arcopagia (Johnsonella) remies	Tellinidae
Tellina scobinata	Arcopagia (Scutarcopagia) scobinata—type	Tellinidae
Tellina lactea	Loripes (Loripes) lactea	Lucinidae
Tellina carnaria	Strigilla (Strigilla) carnaria—type	Tellinidae
Tellina bimaculata	Heterodonax bimaculata	Psammobiidae
Tellina balthica	Macoma (Macoma) balthica	Tellinidae
Tellina pisiformis	Strigilla (Strigillella) pisiformis—type	Tellinidae
Tellina divaricata	Divaricella (Divaricella) divaricata	Lucinidae
Tellina digitaria	Astarte (Digitaria) digitaria type	Astartidae
Tellina cornea	Sphaerium (Sphaerium) cornea—type	Sphaeriidae

During the decades immediately succeeding the publication of the tenth edition of the Systema Naturae, students continued to interpret the genus *Tellina* in the broad sense that it had been employed by Linneaus, and a rather extensive list of species of *Tellina* (sensu lato) was built up. The first to attempt any refinement was Scopoli, who in 1777 created the genus *Sphaerium* for *"Tellina"* *cornea*, a freshwater form and the last species on the Linnaean list.

Chemnitz in 1782, in a work that is, unfortunately, non-binomial, erected the genus *Tellinula;* this was the first proposal of a supra-specific name for forms that would today fall within the limits of the family Tellinidae as at present defined. In 1791 Poli proposed the terms *Peronaea* and *Peronaeaderma* for the animal and the shell, respectively, of the form that Linnaeus had named *Tellina planata*. Although Poli introduced a dual system of nomenclature, his names are available for modern taxonomic purposes and his name *Peronaea* is here considered as a valid member of the family Tellinidae.

Bruguiere in 1797 used the generic names *Cyclas*, *Lucina* and *Capsa* at the top of plates illustrating, but not naming, a number of nontellinid forms that had been referred to the genus *Tellina* by earlier post-Linnaean students. These names were adopted by Lamarck in 1799, gained wide acceptance, and the first two were made the type species of separate families by Fleming in 1828.

In 1817 Schumacher separated four groups from within the heterogenous assemblage of forms that constituted the genus *Tellina* as understood at that time. These he named *Gari*, *Gastrana*, *Phylloda* and *Omala*. *Gari* has subsequently been made the type genus of the family Gariidae; *Gastrana*, *Phylloda* and *Omala* are still to be referred to the Tallinidae as separate genera.

During the following year Lamarck, apparently unaware of Schumacher's work, proposed the name *Psammobia* for the *Tellina gari* of Linnaeus. This is an objective synonym of *Gari* Schumacher. At the same time he also proposed the names *Psammotaea* and *Tellinides* for tellinid species; the former is now held to be a member of the Garidae, the latter is still retained within the Tallinidae, being here considered a subgenus of *Tellina* sensu stricto.

Shortly afterward, in 1819, Leach erected the genus *Macoma* for those tellinids lacking anterior laterals, and in 1827 Brown described the genus *Arcopagia* from a manuscript of Leach proposing the name for the orbicular forms in which the pallial sinus is free from the pallial line. The types of neither of these genera were on the original Linnaean list of species of *Tellina*.

In 1822 Turton separated, under the generic name *Strigilla*, those orbicular forms that are marked by divaricant oblique sculpture.

The other groups that are here recognized as being of generic status were not described until many years later. Thus the genus *Tellidora*, although proposed by Mörch in 1853, was not formally described until 1856 in the classic *Genera of Recent Mollusca* by the brothers, H. and A. Adams. At the same time these authors gave the name *Metis* as a subgeneric name under the genus *Tellina* to one of the forms illustrated under the name *Capsa* by Bruguière in 1797. This species, which had been named *Tellina meyeri* by Dunker, was not conspecific with that which had been selected as type of *Capsa*, namely, *Venus deflorata* Linnaeus, by subsequent designation, Schmidt 1818. This species is also the type of *Asaphis* Modeer, 1793, and *Capsa* falls into the synonymy of that genus. Unfortunately, the name *Metis* was a homonym, having previously been used by Philippi in 1843; the substitute name *Apolymetis* was provided by Salisbury in 1929.

The pre-Linnaean name Quadrans of Klein, 1753, was used and validated by Bertin in his "Revision des Tellinidés" of 1878. Bertin assigned to this genus a number of more or less trigonal forms, naming *Tellina gargadia* the first species on the Linnaean list as type.

These then are the forms that are recognized in this study as possessing generic status in the family Tellinidae: *Tellina* Linnaeus, 1758; *Gastrana*, *Phylloda* and *Omala* Schumacher, 1817; *Macoma* Leach, 1819; *Strigilla* Turton, 1822; *Arcopagia* (Leach) Brown, 1827; *Tellidora* Mörch, 1852; *Quadrans* (Klein) Bertin, 1878; *Apolymetis* Salisbury, 1929; *Linearia* Conrad, 1860 (*see below*); *Finlayella* Laws, 1933; *Angulus* Mühlfeld, 1811; *Barytellina* Marwick, 1924; *Macalia* H. Adams, 1860.

Supraspecific Names Based Upon Fossil Types

During all of this period, such fossil species as were being described were, almost without exception, uncritically referred to the genus *Tellina*, sensu lato. The use of the term *Tellinites* Schlotheim 1820, does not enter into consideration here, because it was proposed as an inclusive term for Paleozoic forms of *Tellina*-like shapes, none of which are referable to the family Tellinidae. It was not until 1860 that the first efforts were made to recognize generic entities within the assemblage of fossil Tellinidae. At this time Conrad proposed the names *Linearia* and *Tellinimera* for American Upper Cretaceous species. In 1870 he added the name *Aenona*, also for an Upper Cretaceous species, and in the same year Stoliczka proposed *Palaeomoera* for *Tellina strigata* Goldfuss, a European Cretaceous form. In 1871 Meek proposed *Arcopagella* for a Cretaceous species from the "Upper Missouri Country," and Conrad in 1875 added the names *Hercodon*, *Liothyris*, and *Oene* for Cretaceous species from North Carolina. In 1944 Olsson added *Tellipiura* to the list of Cretaceous names.

With all of this plethora of names for Cretaceous species during the period 1860 to 1875, it was not until 1886 that superspecific names were proposed for Tertiary fossils. At that time Cossmann described, as sections of the genus *Tellina*: *Elliptotellina*, *Macaliopsis*, *Cyclotellina*, and *Arcopagiopsis*, all being based upon Eocene species from the Paris Basin of France. In 1891 he added the names *Herouvalia* and in 1921 *Sinuosipagia*, both for Eocene species. Pilsbry and Olsson in 1941 proposed *Macoploma* with *Macoma ecuadoriana* as the type species from Ecuador; in 1941 Olsson proposed *Panacoma* with *Macoma chiriquiensis*, a species from Panama, as the type. Marwick proposed *Ascitellina* in 1928 and *Bartrumia* in 1934, both from Lower Miocene deposits of New Zealand.

The first name based upon a post-Eocene fossil type, however, was *Macopsis* Sacco, 1901, of which the type was the Italian Pliocene species *Tellina elliptica* Brocchi. In 1924 Marwick proposed *Barytellina* with the type *B. crassidens*, a Pliocene species from New Zealand. In 1933 Laws described *Finlayella* with type *F. sinuaris*. The name *Oudardia* Monterosato 1884 might well be included in the list, since the type *Tellina oudardii* Payraudeau, described from the Recent

9

fauna of the Mediterranean Sea, has proven to be a synonym of *Tellina compressa* Brocchi described from the Italian Pliocene.

In the course of this work, I came across a number of species which appeared to possess distinctive characteristics, hence requiring designation as the types of certain superspecific categories within the framework of some of the generic groups. Of these, *Tellina staurella* Lamarck is designated type of a new subgenus *Liotribella; Tellina rostrata* Linnaeus is designated type of a new subgenus *Dallitellina* in the honor of W. H. Dall, former Curator of the Division of Mollusks of the U. S. National Museum; *Tellina pulcherrima* Sowerby is designated as the type of a new subgenus *Smithsonella* in honor of James Smithson, founder of the Smithsonian Institution—all three of these subgenera are within the genus *Tellina* Linnaeus.

Within the genus *Arcopagia, Tellina fausta* Pultney is designated type of a new *Johnsonella* which is named in honor of the late C. W. Johnson of the Boston Society of Natural History; *Tellina linguafelis* Linnaeus is designated type of a new subgenus *Smitharcopagia* in honor of E. A. Smith of the British Museum. *Tellina stephensoni* Salisbury is designated type of a new subgenus *Iredalesta* within the genus *Linearia* Conrad, and it is named in honor of Dr. T. Iredale of the Australian Museum, Sydney, Australia.

Within the genus *Strigilla* Turton, I have proposed the subgenus *Roemerilla*, with *Tellina cicercula* Philippi as the type, in honor of Eduard Römer, a 19th century German malacologist. For the species of large size within the genus *Angulus* a new subgenus *Megangulus* is established with *Tellina venulosa* Schrenck as the type.

Tellina gaimardi Iredale is designated type of a new subgenus *Bartschicoma* in honor of the late Dr. Paul Bartsch of the National Museum in the genus *Macoma* Leach. Within the genus *Apolymetis* Salisbury, a new subgenus *Pilsbrymetis*, with *Tellina lacunosa* Schröter as the type, is erected in honor of the late Dr. H. A. Pilsbry of the Academy of Natural Sciences of Philadelphia.

The Classification of the Tellinidae

Perhaps the earliest attempt to present a classification of the large number of species, and considerable number of superspecific names that had accumulated within the family Tellinidae, was published in 1856 by H. and A. Adams in their monumental work *The Genera of Recent Mollusca*, vol. 2. At that time they included within the family a number of genera and subgenera that are no longer considered to represent tellinid types. Their classification may be summarized as follows:

Family Tellinidae
  Subfamily Tellininae
    Genus *Asaphis* Modeer (now referred to Gariidae)
    Genus *Gari* Schumacher (now referred to Gariidae)
      Subgenus *Psammocola* Blainville
      Subgenus *Amphichaena* Philippi
    Genus *Sanguinolaria* Lamarck (now referred to Senguiolariidae)
    Genus *Hiatula* Modeer (now referred to Gariidae)
      Subgenus *Psammotaea* Lamarck
      Subgenus *Psammotella* Deshayes
    Genus *Elizia* Gray (now referred to Gariidae)
    Genus *Tellina* Linnaeus
      Subgenus *Tellinella* Gray
      Subgenus *Peronaeoderma* Mörch
      Subgenus *Moera* H. and A. Adams
      Subgenus *Arcopagia* Leach
      Subgenus *Phylloda* Schumacher
      Subgenus *Angulus* Mühlfeldt
      Subgenus *Tellinides* Lamarck
      Subgenus *Homala* Mörch
      Subgenus *Peronaea* Poli
      Subgenus *Metis* H. and A. Adams
    Genus *Strigilla* Turton
    Genus *Macoma* Leach
    Genus *Tellidora* Mörch
    Genus *Gastrana* Schumacher
    Genus *Lucinopsis* Forbes and Hanley (now referred to Petricolidae)
  Subfamily Donacinae

This classification was confined to the Recent species and there was no apparent attempt to consider the fossil types. So far as can be determined, the characteristics of the dentition were given primary consideration in the arrangement proposed.

The next serious attempt to propose a classification of the Tellinidae was that of Ferdinand Stoliczka who in 1870, in a classical study of the Cretaceous Pelecypoda of India (Geological Survey of India, Memoirs: *Paleontologica Indica*, vol. III, 1870–71), attempted a revision and classification of all of the genera of Pelecypoda described up to that time. The classification of the family Tellinidae that he adopted included three subfamilies: Gariinae, Tellininae, and Capsinae. The first and third of these are no longer referred to the family Tellinidae and need not be considered here; in his subfamily Tellininae he considered the following groups:

Sub-family Tellininae
  Genus *Macoma* Leach, 1819
  Genus *Tellidora* Mörch, 1851 (?)
  Genus *Tellina* Linnaeus, 1758
    Subgenus *Tellinella* Gray, 1852
    Subgenus *Peronaeoderma* Poli, 1795
    Subgenus *Moera* H. and A. Adams, 1852
    Subgenus *Palaeomoera* Stoliczka, 1870
    Subgenus *Arcopagia* Leach, 1827 (teste Brown)
    Subgenus *Linearia* Conrad, 1860
    Subgenus *Phylloda* Schumacher, 1817
    Subgenus *Angulus* Mühlfeldt, 1811 (*Tellinula* Chem., *Fabulina*, Gray)
    Subgenus *Tellinimera* Conrad, 1860
    Subgenus *Tellinides* Lamarck, 1818
    Subgenus *Homalina* Stoliczka, 1870 (*Homala* auct. H. and A. Adams)
    Subgenus *Peronaea* Poli, 1791
    Subgenus *Metis* H. and A. Adams
  Genus *Mactromya* Agassiz, 1842
  Genus *Strigillina* Turton, 1822 (err. pro *Strigilla* Turton)

The generic name *Mactromya* Agassiz was included by Stoliczka on the basis of the general shape of the unusually thick shell; the hinge was then unknown, and he was quite uncertain as to its true systematic position. It is now no longer included within the Tellinidae. The subfamily Capsinae included, among others, the genera *Macalia* H. Adams 1860 and *Gastrana* Schumacher 1817, both of which are now referred to the Tellinidae. Following his classification, Stoliczka made an effort to assign to their proper position within it all of the described fossil Cretaceous species. Thus we have here the first attempt to propose a classification of the Tellinidae based upon characters that were available to the paleontologist and the first effort to assign correctly to a classification any of the fossil species that had previously been described.

In the following year Römer, in a beautifully illustrated report based solely upon the Recent fauna, recognizes only one genus, *Tellina*, with the following five subgenera: *Strigilla* Turton, *Tellidora* Mörch, *Metis* H. and A. Adams, *Macoma* Leach, and *Gastrana* Schumacher. This work, which was entitled "Die Familie der Tellmuscheln, Tellinidae," appeared as the tenth volume of the

Systematisches Conchylien Cabinet; it is of value primarily for the excellent illustrations that it contains.

In 1878 M. V. Bertin published his revision of the Tellinidae (Bertin, 1878). In this work his classification is erected around the following eight genera: *Tellina* Linnaeus, *Strigilla* Turton, *Arcopagia* (Leach) Brown, *Tellidora* Mörch, *Phylloda* Schumacher, *Metis* H. and A. Adams, *Macoma* Leach (in Brown), and *Gastrana* Schumacher. Here, in a classification that appears to be based primarily upon the nature of the dentition, we find for the first time *Arcopagia*, *Metis* and *Phylloda*, considered as distinct genera in a classification of the Tellinidae as a unit.

During the course of his comprehensive studies in the preparation of his well-known monograph modestly titled "Contributions to the Tertiary Fauna of Florida" (Dall, 1900b), Dall made a careful study of the family Tellinidae giving in that work, and in a preliminary report entitled "Synopsis of the Family Tellinidae and of the North American Species" (Dall, 1900a) a classification that was the most complete yet issued. Unfortunately, he considered that Lamarck's citation of *Tellina virgata* Linnaeus, as an example to illustrate the genus, constituted a type designation and was therefore misled with respect to his synonymy of and within the genus *Tellina*. The classification that he proposed in his "Synopsis" is given in the following summary in which I have marked those names with an asterisk (*) that are newly proposed in this classification:

Genus *Tellina* "(Linnaeus) Lamarck," 1799
  Subgenus *Liotellina* Fischer, 1887
    Section *Macaliopsis* Cossmann, 1886
    Section *Herouvalia* Cossmann, 1892
    Section *Arcopagella* Meek, 1871
  Subgenus *Linearia* Conrad, 1860
  Subgenus *Elliptotellina* Cossmann, 1866
  Subgenus *Pseudarcopagia* Bertin, 1878
  Subgenus *Arcopagia* (Leach), 1827
    Section *Cyclotellina* Cossmann, 1886
    Section *Phyllodina* Dall, 1900*
    Section *Merisca* Dall, 1900*
    Section *Eurytellina* Fischer, 1887
    Section *Scrobiculina* Dall, 1900*
    Section *Quadrans* Bertin, 1876
    Section *Tellinides* Lamarck, 1818
  Subgenus *Phylloda* Schumacher, 1817
  Subgenus *Moerella* Fischer, 1887
  Subgenus *Angulus* (Megerle em.), 1811
    Section *Angulus* s.s.
    Section *Scissula* Dall, 1900*
    Section *Oudardia* Monterosato, 1884
    Section *Peronidia* Dall, 1900*
  Subgenus *Omala* Schumacher, 1817
  ? Section *Homalina* Stoliczka, 1871
Genus *Strigilla* Turton, 1822
Genus *Tellidora* (Mörch), 1856

Genus *Metis* H. and A. Adams, 1856
Genus *Gastrana* Schumacher, 1817
Genus *Macoma* Leach, 1819
  Subgenus *Macoma* Leach, 1819, s.s.
    Section *Macalia* H. Adams, 1860
  ? Section *Rexithaerus* Conrad, 1869
  Subgenus *Cymatoica* Dall, 1889
  Subgenus *Psammacoma* Dall, 1900*
    Section *Psammacoma* Dall s.s.
    Section *Cydippina* Dall, 1900*
    Section *Psammotreta* Dall, 1900*

In the more complete statement of his classification published later in the same year in Part 5 of the "Contributions to the Tertiary Fauna of Florida," he removed his question marks as to the status to be accorded *Homalina* Stoliczka and *Rexithaerus* Conrad. Furthermore, under the genus *Strigilla* Turton, he adds the following "Sections": *Strigilla* s.s., *Rombergia* Dall, 1900*, and *Aeretica* Dall, 1900*. Thus, during the course of the classification, ten new groups of supraspecific rank were added to the large list already existing within the family.

Although this classification is based primarily upon the characters of the dentition, there are certain aspects that are very difficult to comprehend on this basis. Thus *Eurytellina* Fischer and *Tellinides* Lamarck, which he has made sections of *Arcopagia*, are far removed from it in the character of dentition, pallial sinus and shape. *Quadrans* of Bertin, which Dall has made another section of *Arcopagia*, possesses enough distinctive characters to form a separate genus, and *Phyllodina* is closer to *Quadrans* than it is to *Arcopagia*. Dall's (1900b) table of classification of genera, which is based on the presence or absence of lateral teeth, places *Apolymetis* in the group possessing laterals, whereas the type of *Apolymetis* is without laterals. Dall (1900b, p. 1015) has put *Homalina* under *Omala* stating that: "Shell resembling *Omala*, but according to literature without any lateral teeth." But the type of *Homalina* has two distinct right laterals and has other morphologic differences from *Omala* Schumacher; therefore, these two could not be grouped together. It was these and other similar factors, together with Dall's misinterpretation of the designated type of the genus *Tellina* sensu stricto, that led to the present attempt at a revision of the family.

In 1934 Thiele, in the third volume of his *Handbuch der Systematischen Weichtierkunde*, gives a classification of the family Tellinidae in which he also includes certain, but not all, of the names that have been proposed on the basis of fossil types. Unfortunately, citation of *Tellina virgata* has been accepted here as a type designation. In this classification fifteen genera are recognized as follows: *Arcopagia* (Leach) Brown, 1827; *Strigilla* Turton, 1822; *Pseudarcopagia* Bertin, 1878; *Cyclotellina* Cossmann, 1886; *Apolymetis* Salisbury, 1929; *Gastrana* Schumacher, 1817; *Macoma* Leach, 1819, *Tellidora* H. and A. Adams, 1856; *Merisca* Dall, 1900; *Quadrans* Bertin, 1878; *Homalina* Stoliczka, 1871; *Phylloda* Schumacher, 1817; *Eurytellina* Fischer, 1887; *Angulus* Mühlfeldt, 1811; *Tellina* Linne, 1758.

It is difficult to understand the basis upon which this classification has been established because Thiele erects distinct genera out of the types that are closely related. Thus we find *Eurytellina* has been given a generic rank, whereas being closely related to *Tellina* Linnaeus, it should be included within that genus. *Pseudarcopagia* Bertin, *Merisca* Dall, and *Cyclotellina* Cossmann, which he has made separate genera, are closely related to each other and to the genus *Arcopagia;* splitting them does not serve the purpose of systematics because their interrelation is lost by doing that. Salisbury, in a general discussion of the Tellinidae published also in 1934 but subsequent to the appearance of Thiele's work, very properly reduces a number of Thiele's "genera" to subgeneric rank adding that unless this be done "no certain line can be drawn anywhere in the Tellinidae."

This review would seem to demonstrate that there is still much confusion as to the proper classification of the family Tellinidae. It is hoped that the present study, being based wholly upon a careful examination of type species, has resulted in a general classification that will be both useful as well as a more natural association of the supraspecific groups within this family.

As has been noted in the previous brief review of the more important efforts to provide a working classification of the family Tellinidae, there has been considerable difference of opinion as to the status to be accorded to a number of the supraspecific units that have been proposed. This appears to stem largely from different concepts as to the basic characters to be used in such a classification of this group.

Unfortunately, most workers have not clearly outlined the basis for their arrangement, and it is often difficult, therefore, to comprehend the reason for some of the assignments made. It has already been stated that the classification here proposed is to be based on those morphologic characters of the shells that are available and useful for paleontologic research. It, therefore, seems desirable to include here a statement of the general philosophy as well as of the specific morphologic characters that has guided the following classification.

In classifying any group of closely related animals, the individual species units within that group are the most significant elements; the larger categories of genus and subgenus are to be erected upon the species and will stand or fall according to the validity of the species-unit upon which they are based.

A species in nature is a group of organisms of like characters; it is not a process, as some geneticists maintain, nor is it a collection of individual specimens as some taxonomists seem unconsciously to assume. While it is true that the group arises as a result of dynamic genetic processes, and that it is composed of individuals, nevertheless, it is the group itself that constitutes the species.

On the other hand, it must be recognized that a species may be differently defined according to the differing qualifications and methods applied in different types of classification. Thus genetic species, morphological species, or taxonomic species will be recognized by the geneticist, the malacologist and paleontologist, or by the taxonomist, respectively, according to the concepts that they apply in their work. A group of organisms in nature having hereditary characters which may be transmitted through reproduction to another succeeding member of the group constitutes a genetic species. However, in the field of paleontology, it is obviously impossible to test the potentiality of transmitting hereditary characters, and the species must be defined morphologically.

In ultimate concept, a morphological species is considered as representing a group of individual animals that possess similar visible morphologic characters such that adjacent local populations within the group differ only in variable characters that intergrade marginally. Such a morphologically defined species is admittedly not a true species in the genetic sense, although it may correspond very closely to a genetic species. This must not be taken to imply that the morphological species is a wholly artificial concept; within the limits of the visible characters, it is a true natural group.

In actual practice, however, the species of the paleontologist is a subjective concept that is constituted as a result of observations made upon a series of individual specimens. This series of available specimens is but a sample drawn from a natural population and may or may not be completely representative of the morphologic characteristics of the entire population. Therefore, the process of defining a species involves the inference from the sample about the characters and limits of the morphological species from which the sample was drawn. The species thus described will, as a result, be a subjective concept rather than the truly objective group that constitutes the true morphologic unit; the degree to which the described species will approximate the ideal morphological species will depend upon the adequacy of the sample and the skill of the paleontologist. The morphological species, as actually constituted, is therefore an inference as to the most probable characters and limits of the true morphological species from which the available sample has been drawn.

The taxonomic species is, and must remain, a purely nomenclatural concept based on the generic and trivial name that has been applied to the zoological species, whether that species be based upon genetic or morphologic considerations.

The basic element of a phylogenetic series is a kind of genetic continuity, actual or potential; that is, the hereditary characteristics are being, or have been in the past, transferable from one part of the population to another. By accepted definition such genetic transfer cannot normally occur between different species, although it is recognized that forms which are genetically incompatible may have arisen, by normal processes of speciation, from ancestral types that were genetically compatible. The philosophical idea around which groups of super-specific status are developed would imply that within the subgenus, genus or family units that, although they are at present genetically distinctive, may be traced, phylogenetically, to common, genetically compatible ancestors. All of the higher taxonomic units are alike in this respect, and the rank to be accorded them is based upon an effort to evaluate the extent to which they have changed under normal evolutionary processes from the ancestral type.

Thus there is in theory an absolute distinction between the species and the genus, but there is no theoretical qualitative difference between the genus and the family. Generic and family categories intergrade, and it is a matter of opinion and convenience as to where the line separating the two be drawn.

According to universal agreement, a genus is one of the lower categories of superspecific rank. It includes either a cluster of species of not distant antecedent common origin, or a single species that differs as much from other known species as if it were more or less central in a cluster from which the other known members are missing.

The genus, thus considered, is one of a number of continuously intergrading taxonomic categories and is therefore incapable of precise definition. In the classification of these superspecific categories the following criteria seem to be most useful: morphology, ontogeny, chronogenesis, and ecology. Of these, the morphology provides the most readily determinable criterion and is therefore the most satisfactory, whereas the other criteria provide corroborative factors useful in checking the conclusions reached from the examination of the morphologic features.

Morphologic characters may be divided into two categories: (a) those characters that prove to be stable and persistent within the group, and (b) those characters that prove to be unstable and subject to variation. In the present classification of the family Tellinidae the following stable characters have been utilized as the basis for the generic classification: (1) the character of the dentition, that is, whether it consists of both cardinals and laterals, or of cardinals alone; (2) the relative nature and position of the teeth with respect to each other and to the elements of the hinge structure; (3) the nature of the pallial sinus, whether it be free or confluent with the pallial line, and its size; the latter factor is considered as being an indication of the nature of the siphons and therefore of the soft anatomy of the animal that secreted the shell.

The less stable characters have been considered as criteria in the classification of infrageneric categories. These include: (1) the variation in the size of the cardinals, including the degree of bifidity of certain of the cardinal teeth; (2) variation in the lateral dental elements, whether they are large, small, or obsolete; (3) variations in shape and relative position of the pallial sinus; (4) the character of the ligamental area, whether it be long or short, deep or shallow; (5) size and shape of the shell; and (6) the nature of the surface sculpture and ornamentation.

The word "type" following a listing indicates that species concerned is the type species of the subgenus indicated; where the subgeneric name is the same as the generic name, it indicates that the species is the type of the genus. On the basis of the above considerations, the following taxonomic classification of the family Tellinidae is here proposed:

PROPOSED CLASSIFICATION OF FAMILY TELLINIDAE

Genus *Tellina* Linnaeus, 1758

Subgenus *Tellina* sensu stricto	Type: *Tellina radiata* Linnaeus
Subgenus *Tellinella* Mörch, 1853	Type: *Tellina virgata* Linnaeus
Subgenus *Liotribella* subg. nov.	Type: *Tellina staurella* Lamarck
Subgenus *Eurytellina* Fischer, 1887	Type: *Tellina punicea* Born
Subgenus *Maoritellina* Finlay, 1926	Type: *Tellina charlottae* Smith
Subgenus *Phyllodella* Hertlein and Strong, 1949	Type: *Tellina insculpta* Hanley
Subgenus *Tellinides* Lamarck, 1818	Type: *Tellina timorensis* Lamarck

Subgenus *Tellinota* Iredale, 1936   Type: *Tellina roseola* Iredale

Subgenus *Hertellina* Olsson 1961   Type: *Tellina (Scissula) nicoyana* Hertlein and Strong

Subgenus *Tellinidella* Hertlein and Strong, 1949   Type: *Telliniaes purpureus* Broderip and Sowerby

Subgenus *Peronaea* Mórch, 1791   Type: *Tellina planata* Linnaeus

Subgenus *Scrobiculina* Dall, 1900   Type: *Tellina viridotincta* Carpenter

Subgenus *Laciolina* Iredale, 1937   Type: *Tellina quoyi* Reeve

Subgenus *Moerella* Fischer, 1887   Type: *Tellina donacina* Linnaeus

Subgenus *Elpidollina* Olsson 1961   Type: *Tellina decumbens* (Carpenter)

Subgenus *Cadella* Dall, Bartsch and Rehder, 1938   Type: *Tellina lechriogramma* Melvill

Subgenus *Elliptotellina* Cossmann, 1886   Type: *Tellina tellinella* Lamarck

Subgenus *Herouvalia* Cossmann, 1892   Type: *Tellina semitexta* Cossmann

Subgenus *Oudardia* Monterosato, 1884   Type: *Tellina compressa* Brocchi

Subgenus *Fabulina* Gray, 1851   Type: *Tellina fabula* Gronovius

Subgenus *Homalina* Stoliczka, 1870   Type: *Tellina triangularis* Chemnitz (=*Tellina trilatera* Gmelin)

Subgenus *Pharaonella* Lamy, 1918   Type: *Tellina pharaonis* Hanley

Subgenus *Dallitellina* new subgenus   Type: *Tellina rostrata* Linnaeus

Subgenus *Smithsonella* new subgenus   Type: *Tellina pulcherrima* Sowerby

Genus *Quadrans* Bertin, 1878

Subgenus *Quadrans* sensu stricto   Type: *Tellina gargadia* Linnaeus

Subgenus *Pistris* Thiele, 1934   Type: *Tellina pristis* Lamarck

Subgenus *Serratina* Pallary, 1920   Type: *Tellina serrata* Brocchi

Subgenus *Phyllodina* Dall, 1900   Type: *Tellina squamifera* Deshayes

Subgenus *Quidnipagus* Iredale, 1929   Type: *Cochlea palatam* Martyn (=*Q. palatam* Iredale)

Subgenus *Pristipagia* Iredale, 1936   Type: *Pristipagia gemonia* Iredale

Subgenus *Acorylus* Olsson and Harbison, 1953   Type: *Tellina (Moerella) suberis* Dall

Subgenus *Obtellina* Iredale, 1929   Type: *Tellina bougei* Sowerby

Genus *Bathytellina* Habe, 1958   Type: *Bathytellina citrocarnea* Kuroda and Habe

Genus *Arcopagia* T. Brown, 1827

Subgenus *Arcopagia* sensu stricto   Type: *Tellina crassa* Pennant

Subgenus *Arcopaginula* Lamy, 1918   Type: *Tellina inflata* Gmelin

Subgenus *Johnsonella* new subgenus   Type: *Tellina fausta* Pultney

Subgenus *Scutarcopagia* Pilsbry, 1918   Type: *Tellina scobinata* Linnaeus

Subgenus *Smitharcopagia* new subgenus   Type: *Tellina linguafelis* Linnaeus

Subgenus *Macaliopsis* Cossmann, 1886   Type: *Tellina barrandei* Deshayes

Subgenus *Hemimetis* Thiele, 1934   Type: *Tellina plicata* Valenciennes

Subgenus *Sinuosipagia* Cossmann, 1921   Type: *Tellina colpodes* Bayan

Subgenus *Pseudarcopagia* Bertin, 1878   Type: *Tellina decussata* Lamarck (=*Tellina victoriae* Gatliff and Gabriel)

Subgenus *Zearcopagia* Finlay, 1926 — Type: *Tellina disculus* Deshayes

Subgenus *Lyratellina* Olsson, 1961 — Type: *Tellina lyra* Hanley

Subgenus *Macomona* Finlay, 1926 — Type: *Tellina liliana* Iredale

Subgenus *Cyclotellina* Cossmann, 1886 — Type: *Tellina lunulata* Lamarck

Subgenus *Arcopagiopsis* Cossmann, 1886 — Type: *Tellina pustula* Deshayes

Subgenus *Arcopella* Thiele, 1934 — Type: *Tellina balaustina* Linnaeus

Subgenus *Clathrotellina* Thiele, 1934 — Type: *Tellina pretiosa* Deshayes (= *Tellina pretium* Salisbury)

Subgenus *Merisca* Dall, 1900 — Type: *Tellina crystallina* Wood (= *Tellina cristallina* Spengler)

Subgenus *Pinguitellina* Iredale, 1927 — Type: *Tellina robusta* Hanley

Subgenus *Punipagia* Iredale, 1930 — Type: *Tellina subelliptica* Sowerby (= *Tellina hypelliptica* Salisbury)

Genus *Linearia* Conrad, 1860
Subgenus *Linearia* sensu stricto — Type: *Linearia metastriata* Conrad
Subgenus *Tellinimera* Conrad, 1860 — Type: *Tellina eborea* Conrad
Subgenus *Aenona* Conrad, 1870 — Type: *Tellina eufaulensis* Conrad
Subgenus *Nelltia* Stephenson, 1952 — Type: *Nelltia stenzeli* Stephenson
Subgenus *Palaeomera* Stoliczka, 1870 — Type: *Tellina strigata* Goldfuss
Subgenus *Arcopagella* Meek, 1871 — Type: *Arcopagella mactroides* Meek
Subgenus *Hercodon* Conrad, 1875 — Type: *Hercodon ellipticus* Conrad
Subgenus *Liothyris* Conrad, 1875 — Type: *Linearia carolinensis* Conrad
Subgenus *Oene* Conrad, 1875 — Type: *Oene plana* Conrad
Subgenus *Iredalesta* new subgenus — Type: *Tellina stephensoni* Salisbury
Genus *Strigilla* Turton, 1822
Subgenus *Strigilla* sensu stricto — Type: *Tellina carnaria* Linnaeus
Subgenus *Pisostrigilla* Olsson, 1961 — Type: *Tellina pisiformis* Linnaeus
Subgenus *Simplistrigilla* Olsson, 1961 — Type: *Strigilla strata* Olsson
Subgenus *Rombergia* Dall, 1900 — Type: *Strigilla rombergi* Mörch
Subgenus *Roemerilla* new subgenus — Type: *Strigilla cicercula* Philippi
Subgenus *Aeretica* Dall, 1900 — Type: *Strigilla senegalensis* Hanley
Genus *Finlayella* Laws, 1933 — Type: *Finlayella sinuaris* Laws
Genus *Tellidora* H. and A. Adams, 1856
Subgenus *Tellidora* sensu stricto — Type: *Tellina burneti* Broderip and Sowerby
Subgenus *Tellipiura* Olsson, 1944 — Type: *Tellidora (Tellipiura) peruana* Olsson

Genus *Angulus* Mühlfeldt, 1811
Subgenus *Angulus* sensu stricto — Type: *Tellina lanceolata* Gmelin
Subgenus *Megangulus* new subgenus — Type: *Tellina venulosa* Schrenck
Subgenus *Tellinangulus* Thiele, 1934 — Type: *Angulus (T.) aethiopicus* Jackel and Thiele
Subgenus *Scissula* Dall, 1900 — Type: *Tellina decora* Say (= *Tellina similis* Sowerby)

Genus *Omala* Schumacher, 1817 — Type: *Tellina hyalina* Gmelin
Genus *Phylloda* Schumacher, 1817 — Type: *Tellina foliacea* Linnaeus
Genus *Barytellina* Marwick, 1924
Subgenus *Barytellina* sensu stricto — Type: *Barytellina crassidens* Marwick

Subgenus *Iraqitellina* Dance and
Eames, 1966

Type: *Iraqitellina iraqensis* Dance and
Eames

Genus *Macoma* Leach, 1819

Subgenus *Macoma* sensu stricto

Type: *Tellina calcarea* Gmelin

Subgenus *Austromacoma* Olsson, 1961

Type: *Macoma constricta* (Bruguiere)

Subgenus *Rexithaerus* Tryon, 1869

Type: *Macoma secta* Conrad

Subgenus *Psammacoma* Dall, 1900

Type: *Psammotoea candida* (Lamarck)

Subgenus *Ardeamya* Olsson, 1961

Type: *Tellina columbiensis* Hanley Bertin

Subgenus *Macoploma* Pilsbry and
Olsson, 1941

Type: *Macoma ecuadoriana* Pilsbry and
Olsson

Subgenus *Bendemacoma* Eames, 1957

Type: *Peronaea nigeriensis* Newton

Subgenus *Macomopsis* Sacco, 1901

Type: *Tellina elliptica* Brocchi

Subgenus *Cydippina* Dall, 1900

Type: *Macoma brevifrons* Say

Subgenus *Pseudometis* Lamy, 1918

Type: *Tellina truncata* Jonas
(= *Tellina praerupta* Salisbury)

Subgenus *Psammotreta* Dall, 1900

Type: *Tellina aurora* Hanley

Subgenus *Temnoconcha* Dall, 1921

Type: *Macoma brasiliana* Dall

Subgenus *Psammothalia* Olsson, 1961

Type: *Tellina cognata* C. B. Adams

Subgenus *Scissulina* Dall, 1924

Type: *Tellina dispar* Conrad

Subgenus *Peronidia* Dall, 1900

Type: *Tellina albicans* Gmelin

Subgenus *Bartschicoma* new subgenus

Type: *Tellina gaimardi* Iredale

Subgenus *Rostrimacoma* Salisbury,
1934

Type: *Panopea cancellata* Sowerby

Subgenus *Panacoma* Olsson, 1942

Type: *Macoma (Panacoma)
chiriquiensis* Olsson

Subgenus *Pinguimacoma* Iredale, 1936

Type: *Pinguimacoma hemicilla* Iredale

Subgenus *Jactellina* Iredale, 1929

Type: *Tellina obliquaria* Deshayes

Subgenus *Loxoglypta* Dall, Bartsch
and Rehder, 1939

Type: *Tellina obliquilineata* Conrad

Subgenus *Exotica* Lamy, 1918

Type: *Tellina (Exotica) triradiata*
H. Adams

Subgenus *Salmacoma* Iredale, 1929

Type: *Salmacoma vappa* Iredale

Subgenus *Ascitellina* Marwick, 1928

Type: *Ascitellina donaciformis* Marwick

Subgenus *Bartrumia* Marwick, 1934

Type: *Raeta tenuiplicata* Bartrum

Subgenus *Cymatoica* Dall, 1889

Type: *Tellina undulata* Hanley

Genus *Apolymetis* Salisbury, 1929

Subgenus *Apolymetis* sensu stricto

Type: *Tellina meyeri* Dunker

Subgenus *Leporimetis* Iredale, 1930

Type: *Tellina spectabilis* Hanley

Subgenus *Florimetis* Olsson and
Harbison, 1953

Type: *Tellina intastriata* Say

Subgenus *Tellinimactra* Lamy, 1918

Type: *Tellina edentula* Spengler

Subgenus *Pilsbrymetis* new subgenus

Type: *Tellina lacunosa* Schroter

Genus *Macalia* H. Adams, 1860

Type: *Tellina bruguieri* Hanley

Genus *Gastrana* Schumacher, 1817

Subgenus *Gastrana* sensu stricto

Type: *Tellina fragilis* Linnaeus

Subgenus *Heteromacoma* Habe, 1952

Type: *Tellina irus* Hanley

Family Tellinidae

DESCRIPTION: Gills of eulamellibrachiate type, nonplicate, small, very posterior, the outer limb dorsally directed, sometimes without a reflected lamina, or obsolete; palpi very large, more or less united behind, occasionally the anterior palps with antenna-like projections in front; byssal apparatus obsolete; foot compressed, short not grooved, sometimes capable of being flattened ventrally for use as a fulcrum; mantle margins duplex, with papillose edge open between the pallial sinus and the anterior adductor; siphons long, unequal, naked, retractile, with papillose orifices, the branchial one without a curtain valve; pallial sinus deep, discrepant in the opposite valves. Animal dioecious, marine.

Shell substance cellulo-crystalline, with an inconspicuous periostracum; valves slightly unequal, free, rounded in front, more or less rostrate, oblique and gaping behind, compressed, usually with smooth margins, low beaks and variable, chiefly concentric sculpture; anterior adductor muscle scar larger than posterior, frequently irregular; pedal scar distinct; one or two isolated small scars, made by attachments of the muscles of the mantle, frequently visible near the ventral posterior termination of the pallial sinus; resilium embraced in the ligament, subexternal; hinge area narrow, anterior lateral lamina proximate to cardinals, posterior ones, when present, more distant from the cardinals; cardinal teeth small.

GEOLOGIC RANGE: Lower Cretaceous to Recent.

REMARKS: The family Tellinidae comprises a very interesting group of mollusks in the class Pelecypoda. While preserving their distinct family characters, they live in every latitude from polar to the equatorial waters and have a world-wide distribution.

During the geologic past the members of this family have had a similarly wide distribution, and many fossil species are reported from the Lower Cretaceous through the Pleistocene formations in all parts of the world. Wherever the fossils are found in well-preserved condition they make excellent index fossils due to their distinctive characters.

KEY TO THE GENERA OF FAMILY TELLINIDAE

A. Laterals present.
    a. Two laterals in right valve.

b. Pallial sinus same in both valves.
  c. Pallial sinus close to or touching anterior adductor scar.
    d. Shell elongate-ovate, sculpture of fine concentric ridges or
      striae.. . . . . *Tellina* s.s.
    dd. Shell trigonal-ovate, sculpture of fine concentric ridges
      becoming coarse on the posterior half of shell, postero-
      dorsal margin spinose. . . . . *Quadrans*
  cc. Pallial sinus not close to anterior adductor muscle scar. . .
    e. Pallial sinus ascending, its lower margin entirely free.
      f. Shell orbicular.. . . . . *Arcopagia*
      ff. Shell elongate-ovate.
        (1) Two cardinals and two laterals in each valve. . . . *Linearia*
        (2) Two cardinals in each valve, two laterals only in
          right valve. . . . . *Bathytellina*
    ee. Pallial sinus not ascending, its lower margin entirely
      coalescent with pallial line, oblique sculpture. . . . . . *Strigilla*
  bb. Pallial sinus discrepant in each valve.
    g. Shell ovate, sculpture of fine concentric and radial striae. . *Finlayella*
    gg. Shell trigonal, sculpture of fine concentric ridges, dorsal
      margins spinose. . . . . *Tellidora*
aa. One lateral only in right valve.
  h. Right valve with anterior lateral only, left anterior lateral
    obsolete or absent.
    i. Lower margin of pallial sinus entirely coalescent with pallial
      line.
      j. Pallial sinus generally close to anterior adductor scar. . . *Angulus*
      jj. Pallial sinus not close to anterior adductor scar. . . . . *Omala*
    ii. Only half of ventral margin of pallial sinus coalescent with
      pallial line. . . . . *Phylloda*
  hh. Right valve with posterior lateral only, left posterior lateral
    obsolete.. . . . . *Barytellina*
B. Laterals absent
  a. Pallial sinus discrepant in two valves.
    b. Cardinal teeth small. . . . . *Macoma*
    bb. Cardinal teeth very large. . . . . *Macalia*
  aa. Pallial sinus not discrepant in two valves.
    c. Shell large and orbicular. . . . . *Apolymetis*
    cc. Shell small and trigonal. . . . . *Gastrana*

GENUS TELLINA LINNAEUS, 1758

DIAGNOSIS: Shell of small to large size, elongate-ovate, more or less rostrate posteriorly, slightly inflated or somewhat compressed, of thin to medium thickness, usually
white, often rayed with orange-red or yellow or completely suffused with same colors.
The surface is variously sculptured, usually concentrically. Ligament is external and
situated in an elongate, more or less depressed ligamental groove. Hinge has two
cardinals in each valve, which are rather small and divergent; lateral teeth in right
valve usually strong, those in the left valve subordinate. Interior is white, reddish or
yellow. Pallial sinus is rather large, its ventral margin partly coalescent with pallial
line; adductor muscle impressions are subequal, anterior one more elongated.

Key to the Subgenera of Genus *Tellina*

A. Anterior laterals distant from the cardinals.
  a. Only half of ventral margin of pallial sinus coalescent with pallial line.
    b. Umbones centrally situated.
      c. Sculpture concentric only.
        d. Sculpture of fine concentric striae. . . . . *Tellina* s.s.
        dd. Sculpture of fine concentric ridges. . . . . *Tellinella*
      cc. Concentric sculpture with scales.
        e. Scales on posterior and anterior ends. . . . . *Smithsonella* new subgenus
        ee. Scales only along dorsal margins.. . . . . *Dallitellina* new subgenus
    bb. Umbones situated posterior to mid-length.
      f. Sculpture concentric only.
        g. Sculpture of fine concentric ridges.
          h. Shell with crimson cross on the umbones. . . . . . . *Liotribella* new subgenus
          hh. Shell without crimson cross on the umbones. . . . . *Cadella*
        gg. Sculpture of fine concentric striae. . . . . *Elliptotellina*
      ff. Concentric sculpture with radial striae on both ends. . . . *Herouvalia*
  aa. More than half of pallial sinus coalescent with pallial line.
    i. Umbones centrally situated. . . . . *Pharaonella*
    ii. Umbones situated posterior to mid-length.
      j. Sculpture of fine radial striae.
        k. Pallial sinus large. . . . . *Scrobiculina*
        kk. Pallial sinus small. . . . . *Laciolina*
      jj. Sculpture of fine concentric ridges.
        (1) Ventral margin of shell with slight sinus in the middle. . *Moerella*
        (2) Ventral margin without sinus. . . . . *Elpidollina*
B. Anterior laterals close to cardinals.
  a. Pallial sinus discrepant in two valves. . . . . *Homalina*
  aa. Pallial sinus not discrepant in two valves.
    b. Umbones centrally situated.
      c. Sculpture concentric only.
        d. Anterior end of pallial sinus coalescent with anterior adductor muscle scar. . . . . *Eurytellina*
        dd. Anterior end of pallial sinus not coalescent with anterior adductor muscle scar.
          (1) In the right valve only posterior cardinal is bifid. . . *Tellinota*
          (2) In the right valve both cardinals are bifid. . . . . . *Hertellina*
      cc. Concentric sculpture with scales on posterior area. . . . . *Phyllodella*
    bb. Umbones not centrally situated.
      e. Umbones situated posterior to mid-length.
        f. Sculpture concentric only. . . . . *Maoritellina*
        ff. Sculpture with radial or oblique striae.
          g. Concentric sculpture with radial striae. . . . . *Tellinidella*
          gg. Concentric sculpture with oblique striae. . . . . . . *Oudardia*
      ee. Umbones situated anterior to mid-length.
        h. Concentric sculpture only. . . . . *Tellinides*

hh. Concentric sculpture with radial or oblique striae.
  i. Concentric sculpture with radial striae. . . . . *Peronaea*
  ii. Concentric sculpture with oblique striae. . . . . . . *Fabulina*

Subgenus *Tellina* sensu stricto

Tellina Linnaeus, 1758, p. 674, (For title of original source *see* References Cited)
Musculus Mörch, 1853, p. 3 (not Röding, 1798).
Liotellina Fischer, 1887, p. 1147.
TYPE: *Tellina radiata* Linnaeus (by subsequent designation, Schmidt, 1818).
GEOLOGIC RANGE: It is known only from the Recent marine fauna, but it is so distinctly separated from the other members of the genus *Tellina* as to suggest a considerable period of development; therefore, it appears quite probable that some members of this subgenus may have existed during Pleistocene time.

Tellina (*Tellina*) *radiata* Linnaeus
(Pl. 1, figs. 1–5)

Tellina radiata Linnaeus, 1758, p. 675; Gmelin, 1791, p. 3232; Lamarck, 1818, p. 520; Wood, 1818, p. 18, pl. 4, fig. 26; Hanley, 1846, p. 245, pl. 63, figs. 220, 221; Römer, 1870, p. 5, pl. 1, figs. 4, 8, pl. 4, figs. 1–4.
Tellina unimaculata Lamarck, 1818, p. 520.
DESCRIPTION: Shell of medium to large size, moderately thick, elongate-ovate, slightly inflated, subequilateral. Anterior end is slightly longer than posterior end and evenly rounded; posterior end is somewhat rostrate and bluntly truncate. Beaks are low, dorsal margins are sloping at a moderate and equal angle; ventral margin is scarcely arcuate, with a slight emargination at its middle part and another one at its posterior end. Postero-umbonal fold is obsolete; ligament is sub-external, rather large and situated in a large, lanceolate and somewhat depressed ligamental groove; lunule is small and narrow. Surface is glossy and white, generally colored with purple or rose colored rays, rarely with only suffused faint yellow on the umbonal area and rose color on the apex of the beaks. Sculpture consists of fine concentric growth lines and very fine radial striae. There are two cardinals in each valve; right posterior and left anterior cardinals are large, trigonal and bifid, with a deep socket on either side; right anterior and left posterior cardinals are small, lamellar and simple. Laterals of right valve are prominent, subequidistant, the posterior one is situated below the posterior end of ligamental groove, anterior one is slightly less distant from cardinals; laterals of left valve are obsolete. Between the beak and posterior lateral in both valves the middle part of nymph is turned upward and its edge stands slightly above the dorsal margin. Pallial sinus is large, its dorsal margin is straight, its anterior end acutely rounded and very close to anterior adductor muscle scar, and posterior half of its ventral margin is coalescent with pallial line. Adductor muscle scars are subequal, the anterior one is elongate, posterior one orbicular.
FIGURED SPECIMEN: USNM 73109 (Abaco, Bahamas); length 70.0 mm (100); height 33.2 mm (47.5); thickness 17.2 mm (24.6).
HABITAT: Cape Hatteras, North Carolina, to West Indies, and north coast of South America.

Subgenus *Tellinella* Mörch

Tellinella Mörch, 1853, p. 13.
TYPE: *Tellina virgata* Linnaeus (by subsequent designation, Stoliczka, 1871).
GEOLOGIC RANGE: Upper Cretaceous to Recent.

Tellina (Tellinella) virgata Linnaeus
(Pl. 2, figs. 1–5)

Tellina virgata Linnaeus, 1758, p. 674; Wood, 1815, p. 16, Pl. 3, fig. 3; Lamarck, 1818, p. 521; Hanley, 1846, p. 228, Pl. 63, fig. 212; Römer, 1870, p. 16, Pl. 2, figs. 8, 9, Pl. 7, figs. 1–5.

DESCRIPTION: Shell of medium size, elongate-ovate, moderately thick, slightly inflated, subequilateral. Anterior end is slightly longer than posterior end and broadly rounded; posterior end is narrower than anterior end, bluntly truncate, with acuminate extremity and slightly bent to right. Beaks are low and situated a short distance posterior of mid-length; antero-dorsal margin is straight and has a gentle slope; postero-dorsal margin is slightly concave and has a moderately steep slope; ventral margin is broadly arcuate. Postero-umbonal fold is stronger on the right valve than on the left; ligament is of medium size, subexternal and situated in a slightly depressed ligamental groove; lunule is very small and mostly on the right valve. Surface is colored with concentric bands of red and white or yellowish white together with rays of same colors extending from umbones to margins; beaks are colored with suffused red color. Sculpture consists of sharp concentric ridges which become slightly coarse on dorsal area and fine microscopic radial striae which cross the ridges. There are two cardinals in each valve; right posterior and left anterior cardinals are trigonal, bifid and larger than the other cardinals; right anterior and left posterior cardinals are small, lamellar, and simple. Laterals are inequidistant, the posterior one is situated below distal end of ligamental groove, anterior lateral is slightly less distant from the beaks; laterals of right valve are strong, those of left valve are obsolete. Interior is faint yellow with the exterior coloring generally showing through. Pallial sinus is large, its dorsal margin posteriorly high and anteriorly descending, its anterior end acutely rounded and extending forward to within five millimeters of anterior adductor muscle scar, the posterior half of its ventral margin coalescent with the pallial line. Anterior adductor scar is slightly larger than posterior one and is elongate in shape; posterior one suborbicular.

FIGURED SPECIMEN: USNM 120297 (Indo-Pacific); length, 63.4 mm (100); height, 36.2 mm (57.0); thickness, 11.1 mm (17.5).

HABITAT: From east coast of Africa through East Indies and Ryukyus to Society Islands.

REMARKS: Members of this subgenus occur rather abundantly in the Indo-Pacific waters, and its fossil forms are among the most abundant and well preserved of the Tellinidae. From the point of view of both time range and morphology, this subgenus could be considered as the central type of the family; it possesses closely related morphologic characters with all other genera.

Liotribella, New Subgenus

DIAGNOSIS: Shell of medium size, subtrigonal-ovate, moderately thick, slightly inflated, inequilateral, sub-donaciform. Posterior end is strikingly shorter than anterior end and bluntly truncate; anterior end is broadly rounded. Postero-dorsal margin is more steep than antero-dorsal margin; ventral margin is broadly arcuate; ligament is of medium size and sub-external. Surface is white and generally colored with discontinuous crimson rays; beaks are usually marked by a crimson cross. Sculpture consists of concentric ridges over entire surface with the exception of a smooth area on the posterior part of left valve. There are two cardinals and two laterals in each valve; laterals of left valve are obsolete. Pallial sinus is large and half of its ventral margin is coalescent with pallial line.

TYPE: *Tellina staurella* Lamarck.

GEOLOGIC RANGE: Recent.

Tellina (Liotribella) staurella Lamarck
(Pl. 3, figs. 1–5)

Tellina staurella Lamarck, 1818, p. 522; Delessert, 1841, Pl. 6, fig. 2; Hanley, 1846, p. 229, Pl. 60, fig. 148, Pl. 61, fig. 171, Pl. 66, fig. 261; Reeve, 1866, Pl. 7, figs. 27a, 27b; Römer, 1870, p. 19, Pl. 2, figs. 11–13, Pl. 7, figs. 6–8.

Tellina scalaris Lamarck, 1818, p. 527.

DESCRIPTION: Shell of medium size, subtrigonal-ovate, moderately thick, slightly inflated, inequilateral, sub-donaciform. Anterior end is broadly rounded; posterior end is bluntly truncate, slightly bent to right, narrower and shorter than anterior end. Beaks are situated about 0.36 of the length from posterior end; antero-dorsal margin slopes gently; postero-dorsal margin has very steep slope; ventral margin is broadly arcuate and is slightly sinuous at the posterior part. Postero-umbonal fold is stronger on right valve than on the left. Ligament is of medium size, lanceolate, sub-external and situated in a rather shallow ligamental groove; lunule is represented by a short and very narrow shelf only on the right valve. Surface is colored with discontinuous crimson rays and beaks are marked by a crimson cross. Sculpture consists of sharp concentric ridges over the entire surface with the exception of a smooth area on posterior part of left valve extending from umbone to the posterior end of ventral margin. Right posterior and left anterior cardinals are trigonal, bifid, and larger than the other cardinals; right anterior and left posterior cardinals are small, lamellar, and simple. Laterals are inequidistant: posterior one is situated below the distal end of ligamental groove, anterior lateral is less distant from the beak. Right anterior lateral is strikingly larger than posterior lateral; laterals of left valve are obsolete. Interior of the shell is colored entirely with yellow color except along ventral margin where the crimson rays of the surface show through. Pallial sinus is large; its dorsal margin is nearly horizontal; its anterior end is acutely rounded and extends forward within four millimeters of anterior adductor muscle scar; posterior half of its ventral margin is coalescent with pallial line. Adductor scars are subequal: the anterior one is elongate, posterior one is sub-orbicular.

FIGURED SPECIMEN: USNM 344676 (Ryukyu Islands); length, 55.0 mm (100); height, 33.6 mm (61.0); thickness, 11.4 mm (20.8).

HABITAT: East Africa to Ryukyus, Micronesia, Samoan Islands and Australia.

REMARKS: Lamarck distinguishes three varieties of *T. staurella*. The variety (a) which is white with crimson cross on the umbones, and crimson interrupted rays, I have designated as the type of this section; variety (b) with crimson cross on the umbones, but without rays; variety (c) without umbonal cross and with only faint rays.

T. staurella is distinguished from *T. virgata* in that the former is less elongate, its posterior end is strikingly short, postero-dorsal margin has very steep slope, hinge plate is thin, right anterior lateral is very strong, dorsal margin of pallial sinus is nearly straight, and umbones are generally marked by a crimson cross, and sculpture lacks the very fine radial striae of *virgata;* whereas, the latter is more elongate, subequilateral, its posterior end is subrostrate, the postero-dorsal margin has less steep slope, dorsal margin of pallial sinus convex, the sculpture has very fine radial striae, and umbones are never marked by a crimson cross.

Subgenus *Eurytellina* Fischer

Eurytellina Fischer, 1887, p. 1147.

TYPE: *Tellina punicea* Born (by monotypy)

GEOLOGIC RANGE: The type species is from the Recent fauna of the Caribbean; the fossil forms of this subgenus are represented in the Tertiary fauna, especially in the Miocene of Bowden, Jamaica, and of the Alum Bluff group of Florida.

Tellina (*Eurytellina*) *punicea* Born
(Pl. 3, figs. 6–10)

Tellina punicea Born, 1778, p. 22; 1780, p. 33, Pl. 2, fig. 8; Lamarck, 1818, p. 525 (in part); Hanley, 1847, p. 239, Pl. 60, fig. 154; Römer, 1872, p. 97, Pl. 25, figs. 1–3.
Donax martinicensis Lamarck, 1818, p. 552.

DESCRIPTION: Shell of medium size, rather thick, elongate-ovate, compressed, inequivalve, subequilateral, purple with some concentric white bands. Anterior end is slightly shorter than posterior end and smoothly rounded; posterior end is straight, bluntly truncate, and narrower than anterior end. Beaks are situated slightly anterior of mid-length; postero-umbonal fold is obsolete; and ligament is rather short, lanceolate, subexternal, and situated in a shallow ligamental groove. Dorsal margins are straight and have an equal and steep slope; ventral margin is moderately arcuate and its middle portion is straight. Right posterior and left anterior cardinal teeth are of medium size, trigonal, and bifid; right anterior cardinal is small and trigonal; left posterior cardinal is lamellar, thin, and partly fused with the nymph. Laterals of right valve are moderately prominent, the posterior one is situated below distal end of ligamental groove, anterior one adjacent to cardinals; laterals of left valve are obsolete. Interior has suffused purple color; pallial sinus is large, its dorsal margin is nearly straight, its anterior end is touching anterior adductor muscle scar, its ventral margin is entirely coalescent with pallial line. Adductor muscle scars are subequal, anterior one is elliptically elongate, posterior one is sub-orbicular.

FIGURED SPECIMEN: USNM 181845 (Wounta, Nicaragua); length, 40.1 mm (100); height, 25.0 mm (62.3); thickness, 8.7 mm (21.6).

HABITAT: Cape Hatteras, North Carolina, to northeast coast of Brazil, including Caribbean and Gulf of Mexico.

Subgenus Maoritellina Finlay

Maoritellina Finlay, 1926, p. 466.
TYPE: *Tellina charlottae* Smith (by original designation)
GEOLOGIC RANGE: Recent. Dr. Marwick (1929, p. 913) has described a shell from New Zealand Tertiary as *Maoritellina imbellica*, but his illustration indicates it is closer to *Arcopagia* than to *Maoritellina*.

Tellina (*Maoritellina*) *charlottae* Smith
(Pl. 4, figs. 1–3)

Tellina charlottae Smith, 1885, p. 100, Pl. 4, figs. 1, 1a, 1b.
DESCRIPTION: Shell small, thin, white, compressed, elongae-ovate, inequilateral. Anterior end is long and rounded, posterior end is somewhat narrow, and bluntly truncate. Beaks are small, opisthogyrate, and situated a short distance behind the mid-length; antero-dorsal margin is sloping gently, postero-dorsal margin has steep slope; ventral margin is slightly arcuate. Ligament is small and situated in a shallow ligamental groove, postero-umbonal fold is moderate. Sculpture consists of fine concentric ridges over the entire surface except on the posterior area where it becomes somewhat lamellar. In the original description the cardinals are described as: "two cardinals in right valve and one in left valve, that in latter and posterior in the former being somewhat cleft at the top." I have not seen a specimen of this species. Although

the original illustration is an excellent one, it is difficult to observe details or dentition characteristics; therefore, I will hesitate to change the statement about the number of cardinals in the left valve, but I am inclined to believe that there is also a small, thin, left posterior cardinal partly fused to the nymph as is the case in *T. punicea*. Laterals are slender, elongate, the anterior one rather nearer the cardinals than the posterior. Pallial sinus is large and deep, its anterior end rounded and very close to anterior adductor muscle scar, and its ventral margin is coalescent with pallial line.

FIGURED SPECIMEN: A specimen of *T. charlottae* was not available; therefore the illustration is a copy of the original figure by Smith in "Challenger" report.

HABITAT: New Zealand.

Subgenus *Phyllodella* Hertlein and Strong

Phyllodella Hertlein and Strong, 1949, p. 87.
TYPE: *Tellina insculpta* Hanley (by original designation).
GEOLOGIC RANGE: Recent.

Tellina (Phyllodella) insculpta Hanley
(Pl. 4, fig. 4)

Tellina insculpta Hanley, 1844, p. 70; 1846, p. 289, Pl. 60, fig. 136; Reeve, 1867, Pl. 37, fig. 208.

DESCRIPTION: Shell of medium size, thin, white, compressed, elongate-ovate, equilateral. Anterior end is smoothly rounded; posterior end is straight, bluntly truncate, and slightly narrower than anterior end. Dorsal margins are straight and slope down equally on either side with a moderate angle; ventral margin is only slightly arcuate. Umbones are centrally situated; ligament is small and lanceolate; postero-umbonal fold is moderate. Sculpture consists of fine concentric ridges and some widely spaced growth lines, both crossed by fine radial striae; sculpture of posterior area consists of small scales. Right valve has two bifid cardinals and an anterior lateral situated near the cardinals, and a small posterior lateral; left valve has a bifid anterior cardinal, a thin posterior cardinal and a faint anterior lateral. Pallial sinus is rather high posteriorly and descending anteriorly, the anterior end of it is almost touching the posterior basal margin of the anterior adductor muscle scar, its ventral margin is coalescent with the pallial line.

FIGURED SPECIMEN: The figure shown here is a copy of Hanley's illustration; the locality is given as Chiriqui, west coast of Panama; and the measurements are: length, 33.6 mm (100); height, 18.3 mm (54.5); thickness, 5.3 mm (17.3).

HABITAT: Champerico, Guatemala, to Santa Elena Bay, Ecuador.

Subgenus *Tellinides* Larmarck

Tellinides Lamarck, 1818, p. 535.
TYPE: *Tellina timorensis* Lamarck (by monotypy).
GEOGRAPHIC RANGE: Recent.

Tellina (Tellinides) timorensis Lamarck
(Pl. 4, figs. 5–9)

Tellina timorensis Lamarck, 1818, p. 536; Hanley, 1846, p. 292, Pl. 61, figs. 158, 172; Philippi, 1846, p. 90, Pl. 4, fig. 3; Römer, 1872, p. 166, Pl. 34, figs. 4–6.
Tellina tridentata Anton, 1837, p. 283.

DESCRIPTION: Shell of medium size, white, rather thin, elongate-ovate, compressed, equivalve, inequilateral, slightly gaping at the ends. Anterior end is short and smoothly rounded; posterior end is straight, bluntly truncate, and slightly narrower than anterior end. Dorsal margins are straight and have moderate slope, the postero-dorsal margin is more steep than antero-dorsal margin; ventral margin is scarcely convex and is slightly sinuous at the posterior part. Postero-umbonal fold is obsolete; umbones are situated a short distance anterior of mid-length; ligament is small, lanceolate, sub-external, and situated in a shallow ligamental groove. Sculpture consists of concentric ridges with bands of fine concentric striae in between them. Right posterior and left anterior cardinals are trigonal and bifid; right anterior and left anterior cardinals are small, thin, and lamellar. Right anterior lateral is adjacent to cardinals; right posterior lateral is situated below the distal end of ligamental groove; laterals of left valve are obsolete. Pallial sinus is large, dorsal margin of it is posteriorly high and anteriorly descending, anterior end of it is acutely rounded and extends within two millimeters of anterior adductor muscle scar, ventral margin of it is entirely coalescent with pallial line. Adductor muscle scars are elliptically elongate and subequal.

FIGURED SPECIMEN: USNM 128450 (East Indies); length, 42.4 mm (100); height, 26.4 mm (62.4); thickness, 8.7 mm (20.6).

HABITAT: Pacific and Indian Oceans; Red Sea.

REMARKS: In 1818 Lamarck proposed *Tellinides* as a genus of Tellinidae referring to it only the species *T. timorensis;* however, he misinterpreted the adjacent small anterior lateral of the right valve as a third cardinal tooth. Although he gave no illustration in his work, his description is clear and adequate for identification.

Subgenus *Tellinota* Iredale

Tellinota Iredale, 1936, p. 281

TYPE: *Tellinota roseola* Iredale (by original designation).

GEOLOGIC RANGE: Recent.

Tellina (Tellinota) roseola (Iredale)
(Pl. 4, figs. 10–14)

Tellinota roseola Iredale, 1936, p. 281, Pl. 20, fig. 18.

DESCRIPTION: Shell moderately large, somewhat thin, elongate-ovate, compressed, equilateral, slightly inequivalve. Anterior end is broadly rounded; posterior end is subrostrate, obliquely truncate with acuminate extremity and slightly bent to right; left valve is slightly more convex than right valve. Beaks are rather small and centrally situated; antero-dorsal margin is slightly convex and has a gentle slope, postero-dorsal margin is slightly concave and has a moderately steep slope; ventral margin is broadly arcuate and middle part of it is straight. Postero-umbonal fold is weak, ligament is of medium size, lanceolate and situated in a depressed ligamental groove. Surface is rose-colored with some concentric ridges which become coarse on the posterior part of the shell; middle of the shell up to and including umbonal area is smooth; the concentric sculpture is crossed by very fine almost microscopic radial striae. Hinge plate is thin; right posterior and left anterior cardinals are trigonal, bifid, and with a deep socket on each side; right anterior cardinal is trigonal and small, left posterior cardinal is small, thin, lamellar, and partly fused with the nymph. Right valve has two small laterals; the anterior one is adjacent to cardinals, posterior one is situated below the posterior end of ligamental groove; laterals of left valve are obsolete. Interior of the shell is entirely suffused with rose color. Pallial sinus is large, its dorsal margin is posteriorly high and anteriorly descending, its anterior end is rounded and extends forward within four millimeters of anterior adductor muscle scar, and its ventral margin is entirely coalescent with pallial line. Anterior adductor muscle scar is larger than the posterior adductor scar and is elongate; posterior adductor muscle scar is suborbicular.

FIGURED SPECIMEN: The specimen from the collections of Australian Museum received through the courtesy of Dr. Iredale; length, 53.0 mm (100); height, 31.0 mm (58.5); thickness, 9.6 mm (18.1).

HABITAT: Twofold Bay (locality of the figured specimen) and Byron Bay, both New South Wales.

Subgenus *Tellinidella* Hertlein and Strong

Tellinidella Hertlein and Strong, 1949, p. 79
TYPE: *Tellinides purpureus* Broderip and Sowerby (by original designation).
GEOLOGIC RANGE: Recent.

Tellina (Tellinidella) purpureus (Broderip and Sowerby)
(Pl. 5, figs. 1–5)

Tellinides purpureus Broderip and Sowerby, 1829, p. 363, Fig. 15.

Tellinides purpurascens Broderip and Sowerby, 1839, p. 153, Pl. 42, fig. 2 (not *Tellina purpurascens* Gmelin, 1791); Hanley, 1846, p. 295, Pl. 62, fig. 194; Reeve, 1867, Pl. 20, fig. 103.

DESCRIPTION: Shell of moderately large size, thin, elongate-ovate, compressed, sub-equilateral, both ends slightly gaping. Anterior end is slightly longer than posterior end and sharply rounded; posterior end is obliquely truncate with acuminate extremity. Beaks are low; dorsal margins are sloping at a moderate and equal angle on both sides, antero-dorsal margin is slightly convex, and postero-dorsal margin is straight; ventral margin is broadly arcuate. Postero-umbonal fold is weak, ligament is of medium size, lanceolate, external, and situated in a shallow ligamental groove. Exterior and interior are colored with suffused purplish-rose except for a narrow zone along the margins, which is white. Sculpture consists of fine concentric and radial decussating ridges that become finer toward the umbones. Hinge plate is thin; in the right valve the posterior cardinal is relatively large, trigonal, and bifid, anterior cardinal is small with a deep socket on each side; right anterior lateral is small and adjacent to cardinals, right posterior lateral is obsolete and situated below the posterior end of ligamental groove. In the right valve the anterior cardinal is small and bifid, posterior cardinal is lamellar and obsolete; laterals of left valve are absent. Pallial sinus is large; its dorsal margin is posteriorly high and anteriorly descending, its anterior end is rounded and extends forward within four millimeters of anterior adductor muscle scar, its ventral margin is entirely coalescent with pallial line. Anterior adductor muscle scar is slightly larger than posterior one and has elongate shape, posterior adductor muscle scar is suborbicular.

FIGURED SPECIMEN: USNM 169982 (Corinto, Nicaragua); length, 53.8 mm (100); height, 281 mm (52.2); thickness, 7.9 mm (14.8).

HABITAT: Gulf of California to Northern Peru.

Subgenus *Peronaea* Poli

Peronaea Poli, 1791, p. 29.
TYPE: *Tellina planata* Linnaeus (by subsequent designation Stoliczka, 1870).
GEOLOGIC RANGE: Miocene and Pliocene of France and Italy; Recent.

Tellina (Peronaea) planata Linnaeus
(Pl. 5, figs. 6–10)

Tellina planata Linnaeus, 1758, p. 675; Gmelin, 1791, p. 3232; Lamarck, 1818, p. 525; Römer, 1871, p. 115, Pl. 1, fig. 2, Pl. 28, fig. 1–4.

Tellina operculata var. B Gmelin, 1791, p. 3235.

Tellina complanata Gmelin, 1791, p. 3239.

Omala inaequivalvis Schumacher, 1817, p. 129, Pl. 10, fig. 1.

DESCRIPTION: Shell of medium size, compressed, elongate-ovate, inequivalve, sub-equilateral, white with faint orange color on the umbonal area and brown periostracum around margins. Anterior end is broadly rounded; posterior end has acuminate extremity and is slightly bent to right. Beaks are situated slightly anterior to mid-length; dorsal margins on both sides have a gentle and equal slope; postero-dorsal margin is slightly excavated behind the beaks; ventral margin is broadly arcuate. Postero-umbonal fold is weak on the left valve but rather prominent on the right valve; a short, stout ligament is situated in a deep ligamental groove. Sculpture consists of fine concentric striae which become slightly coarse near margins; these are crossed by fine radial striae. Right posterior and left anterior cardinals are somewhat large, trigonal, bifid, and with a deep socket on each side; right anterior and left posterior cardinals are small and thin. Right valve has two small laterals, the anterior one is adjacent to cardinals, posterior one is situated below posterior end of ligamental groove; laterals of left valve are obsolete. Pallial sinus is large, its dorsal margin is posteriorly high and anteriorly descending, its anterior end is acutely rounded and extends forward to within two millimeters of anterior adductor muscle scar, its ventral margin is entirely coalescent with pallial line. Adductor muscle scars are subequal, the anterior one is elongate, posterior one is suborbicular.

FIGURED SPECIMEN: USNM 178252 (Algiers); length, 51.0 mm (100); height, 33.7 mm (66.1); thickness, 11.4 mm (22.4).

HABITAT: Mediterranean Sea, and the Atlantic coast of Iberian peninsula.

Subgenus *Scrobiculina* Dall

Scrobiculina Dall, 1900, p. 290.

Schumacheria Cossmann, 1902, p. 52.

TYPE: *Scrobicularia viridotincta* Carpenter (by original designation).

GEOLOGIC RANGE: Recent.

Tellina (*Scrobiculina*) *viridotincta* (Carpenter)
(Pl. 6, figs. 1–5)

Scrobicularia viridotincta Carpenter, 1856, p. 160.

Peronaeoderma viriditincta Stearns, 1873, p. 131.

Macoma viriditincta Stearns, 1894, p. 156.

Tellina viridotincta Pilsbry and Lowe, 1932, p. 133.

DESCRIPTION: Shell of medium size, white, rather thin, sub-oval, inequivalve, inequilateral. Anterior end is longer than posterior end and broadly rounded; posterior end is tapering, slightly bent to right, and somewhat gaping. Beaks are small and situated a short distance posterior of mid-length; dorsal margins are sloping at moderate angle, the postero-dorsal margin is slightly steeper than antero-dorsal margin; ventral margin is convex. A medium size ligament is situated in a deep ligamental groove, postero-umbonal fold is weak on the right valve and obsolete on left valve. Sculpture consists of fine concentric ridges and very fine radial striae. Right anterior and left posterior cardinals are small and thin; right posterior and left anterior cardinals are trigonal, bifid, and larger than other cardinals. Right valve has two inequidistant laterals; the posterior one is small and situated below the distal end of ligamental groove, anterior one is well developed and less distant from the cardinals; laterals of left valve are obsolete. Pallial sinus is rather large, its dorsal margin is anteriorly descending, its anterior end is acutely rounded and extends forward to within seven millimeters of anterior adductor muscle scar, and its ventral margin is entirely coalescent with pallial line. Adductor muscle scars are subequal, the anterior one is sub-oval and posterior one is suborbicular.

FIGURED SPECIMEN: USNM 73483 (Gulf of California); length, 55.3 mm (100); height 38.7 mm (70.0); thickness, 12.6 mm (22.8).

HABITAT: Gulf of California to Panama.

REMARKS: In 1902 Cossmann, considering the prior name *Scrobiculinus* Monterosato to be homonymous, substituted *Schumacheria* for *Scrobiculina* Dall 1900; however, this was unnecessary and Dall's name is retained.

Subgenus *Laciolina* Iredale

Laciolina Iredale, 1937, p. 241.

TYPE: *Tellina quoyi* Reeve (by original designation).

GEOLOGIC RANGE: Recent.

Tellina (Laciolina) quoyi Reeve
(Pl. 6, figs. 6–7; Pl. 7, figs. 1–3)

Tellina quoyi Reeve, 1868, Pl. 53, fig. 314; Iredale, 1937, p. 241, Pl. 16, fig. 6.

Tellina lata Hedley, 1913, p. 272.

DESCRIPTION: Shell large, rather thick, white with rose colored rays, elongate-ovate, inequilateral, inequivalve, right valve inflated, left valve somewhat flat. Anterior end is broadly rounded, and slightly bent to left; posterior end is short, sharply pointed, and slightly bent to right. Beaks are small and situated about 0.37 of the length from the posterior end, a medium size ligament is situated in a deep ligamental groove, postero-umbonal fold is obsolete. Dorsal margins form a moderate sigmoidal outline; antero-dorsal margin has a gentle slope, postero-dorsal margin is steep; ventral margin is broadly arcuate and is sinuous at the posterior part. Sculpture consists of fine concentric ridges and very fine radial striae. Hinge plate and cardinal teeth are weak; right anterior and left posterior cardinals are small and lamellar; right posterior and left anterior cardinals are trigonal, bifid, and larger than the other cardinals. Right valve has two subequidistant and well developed laterals, the posterior one is situated below the distal end of ligamental groove, and anterior one is slightly less distant from the beak; laterals of left valve are obsolete. Pallial sinus is small and low, its dorsal margin starts from the lower margin of posterior adductor muscle scar and is descending anteriorly, its anterior end is acutely rounded and extends forward scarcely beyond the middle of the shell, and its ventral margin is almost entirely coalescent with pallial line. Adductor muscle scars are subequal and small, and both have elliptically elongate shape.

FIGURED SPECIMEN: The specimen was obtained from Australian Museum through the courtesy of Dr. Iredale: (Lord Howe Island); length, 86.0 mm (100); height, 55.0 mm (64.0); thickness 23.4 mm (27.2).

HABITAT: Lord Howe Island.

Subgenus *Moerella* Fischer

Moerella Fischer, 1887, p. 1147.

Moera Adams, 1856, non Huebner, 1819.

Maera Adams, 1856, non Leach, 1814.

Donacilla Gray, 1851, non Lamarck, 1812.

TYPE: *Tellina donacina* Linnaeus (by monotypy).

GEOLOGIC RANGE: Lower Miocene to Recent.

Tellina (*Moerella*) *donacina* Linnaeus
(Pl. 8, figs. 1–5)

Tellina donacina Linnaeus, 1758, p. 676; Lamarck, 1818, p. 527; Hanley, 1846, p. 232, Pl. 56, fig. 12; Römer, 1870, p. 26, Pl. 9, figs. 8–12.

Tellina lantivyi Payraudeau, 1827, p. 40, Pl. 1, figs. 13–15.

DESCRIPTION: Shell small, solid, pink with radial, rose coloration, subovate-elongate, somewhat inflated, strongly inequilateral. Anterior end is rounded; posterior end is much shorter than anterior end, bluntly truncate, and slightly bent to right. Beaks are low and situated 0.28 of the length from posterior end; antero-dorsal margin straight and horizontal, postero-dorsal margin steeply sloping; ventral margin horizontal, nearly straight with a slight sinus in the middle. A small ligament is situated in a shallow ligamental groove; postero-umbonal fold is weak on the right valve and absent on the left valve. Sculpture consists of fine concentric ridges and widely spaced growth lines. In the right valve the anterior cardinal is small and lamellar; posterior cardinal is trigonal, bifid, and larger than the other cardinal; laterals are inequidistant, the posterior one is small and situated below the distal end of ligamental groove, anterior one is well developed and less distant from the beaks. In the left valve the anterior cardinal is relatively large and bifid, posterior cardinal is small, lamellar and partly fused with the nymph; laterals of left valve are obsolete. Pallial sinus is large, its dorsal margin is posteriorly high and anteriorly descending, its anterior end is rounded and extends forward very close to the lower margin of anterior adductor muscle scar, its ventral margin is entirely coalescent with pallial line. Anterior adductor muscle scar is slightly larger than posterior one and has elongate shape; posterior one is suborbicular.

FIGURED SPECIMEN: USNM 178334 (Exmouth, England); length, 28.0 mm (100); height, 15.0 mm (53.6); thickness, 9.0 mm (32.2).

HABITAT: North Sea, Atlantic coasts of Great Britain, France, Spain, Portugal and Mediterranean.

REMARKS: In earlier literature some of the species of *Moerella* have been referred to *Angulus;* this confusion was noticed by Dr. Julia Gardner (1928, p. 195) with regard to the east American Miocene and Pliocene species. *Angulus* and *Moerella* may be easily separated in that the former possesses only a right anterior lateral, whereas, *Moerella* has two distinct laterals in the right valve.

Subgenus *Cadella* Dall, Bartsch and Rehder

Cadella Dall, Bartsch and Rehder, 1938, p. 196.

TYPE: *Tellina lechriogramma* Melvill (by original designation).

GEOLOGIC RANGE: Recent.

Tellina (*Cadella*) *lechriogramma* Melvill
(Pl. 8, figs. 6–10)

Tellina lechriogramma Melvill, 1893, p. 14, Pl. 1, fig. 22; Dall, Bartsch and Rehder, 1938, p. 196.

DESCRIPTION: Shell small, ovate-donaciform, somewhat thick, white tinged with rose, rather inflated, strongly inequilateral. Posterior end is very short and bluntly truncate; anterior end is rounded and narrower than posterior end. Beaks are low, opisthogyrate, and situated about a quarter of the length from posterior end. Antero-dorsal margin is sloping very gently; postero-dorsal margin has a very steep slope; ventral margin is straight at the posterior part and broadly arcuate anteriorly. Ligament is small and situated in a depressed ligamental groove; lunule is small, narrow, and situated mostly on the left valve; postero-umbonal fold is obsolete. Sculpture consists of fine concentric ridges which are crowded and sharp on the posterior area. Right anterior and left posterior cardinals are small and lamellar; right posterior and left anterior cardinals are trigonal, bifid, and larger than the other cardinals. Laterals of left value are well developed and the anterior one is more distant from the cardinals than is the posterior one; laterals of left value are obsolete. Pallial line is large and slightly ascending, its ventral margin is partly coalescent with pallial line. Anterior adductor muscle scar is slightly larger than posterior one and elliptically elongate; posterior one suborbicular.

FIGURED SPECIMEN: USNM 130068 (Karachi, Pakistan); length, 14.6 mm (100); height, 9.1 mm (62.4); thickness, 4.6 mm (31.5).

HABITAT: West coast of India. Two species have been described from the Hawaiian Islands.

Subgenus *Elliptotellina* Cossmann

Elliptotellina Cossmann, 1886, p. 70.
TYPE: *Tellina tellinella* Lamarck (by original designation).
GEOLOGIC RANGE: Lower Eocene to Recent.

Tellina (Elliptotellina) tellinella (Lamarck)
(Pl. 8, figs. 11–15)

Donax tellinella Lamarck, 1805, p. 138; Deshayes, 1824, p. 111, Pl. 18, figs. 9, 10, 11.
Tellina exclusa Deshayes, 1860, p. 333.
Tellina subtilis Deshayes, 1860, p. 334, Pl. 25, figs. 15–17.

DESCRIPTION: Shell small, thin, elongate-ovate, rather inflated, inequilateral. Anterior end is slightly longer than posterior end and acutely rounded; posterior end is bluntly truncate and somewhat broader than anterior end. Beaks are low; dorsal margins have gentle slope and ventral margins are scarcely convex. Postero-umbonal fold is obsolete, escutcheon is short and narrow. Sculpture consists of fine concentric lirae. Right anterior and left posterior cardinals are small and lamellar; right posterior and left anterior cardinals are larger than the other cardinals and slightly bifid. Right valve has two inequidistant and well developed laterals, each placed directly above respective muscle scar, thus the anterior lateral is more distant from the cardinals than the posterior one; laterals of the left valve are absent. Pallial sinus is small, ascending, and ventral margin of it is entirely free. Anterior adductor muscle scar is slightly larger than posterior one and ovally elongate; the posterior one is suborbicular.

FIGURED SPECIMEN: USNM 142308 (Grignon, Paris Basin); length, 7.8 mm (100); height, 4.0 mm (51.3); thickness, 2.2 mm (28.2).

OCCURRENCE OF SUBGENUS: Eocene beds of Paris Basin and warm temperate waters of both coasts of North America.

Subgenus *Herouvalia* Cossmann

Herouvalia Cossmann, 1891, in Harris & Burrows, pp. 67, 103, 114; Cossmann, 1892, p. 24.
TYPE: *Herouvalia semitexta* Cossmann (by subsequent designation Cossmann, 1892).
GEOLOGIC RANGE: Lower Eocene.

Tellina (Herouvalia) semitexta (Cossmann)
(Pl. 8, figs. 16–17)

Asaphinella semitexta Cossmann, 1885, p. 99, Pl. 5, figs. 33–35.
Herouvalia semitexta Cossmann, 1892, p. 24, Pl. 1, figs. 15–16.

DESCRIPTION: Shell small, thin, moderately convex, elongate-ovate, equivalve, subequilateral. Posterior end is bluntly truncate; anterior end is acutely rounded and narrower than posterior end. Beaks are somewhat prominent and situated slightly posterior of mid-length. Dorsal margins slope moderately, and ventral margin is only scarcely arcuate. Lunule and escutcheon are narrow, sharply defined, and deep; postero-umbonal fold is obsolete. Sculpture consists mainly of fine concentric ridges with radial striae on both ends which reticulate with the concentric ridges. There are two cardinals and two laterals in each valve; right posterior and left anterior cardinals are bifid and long, the other cardinals are small and separated from bifid cardinals by a deep and triangular socket. Laterals are distant; those of right valve are well developed and parallel to margin, and the ones on the left are obsolete. Pallial sinus is of medium size, ascending; its anterior end is squarish and extending to the middle of the shell; slightly more than half of its ventral margin is free.

FIGURED SPECIMEN: The figures appearing in this work are copies of original illustrations by Cossmann. The measurements are: length, 2.75 mm; height, 1.25 mm.

OCCURRENCE: Herouval, Paris Basin.

Subgenus *Oudardia* Monterosato

Oudardia Monterosato, 1884, p. 22.
TYPE: *Tellina compressa* Brocchi (by original designation).
GEOLOGIC RANGE: Pliocene to Recent.

Tellina (Oudardia) compressa Brocchi
(Pl. 8, figs. 18–22)

Tellina compressa Brocchi, 1814, p. 514, Pl. 12, fig. 9 (not *Macoma compressa* Deshayes, 1854).
Tellina oudardi Payraudeau, 1827, p. 40, Pl. 1, figs. 16–18; Hanley, 1846; p. 297, Pl. 66, fig. 262; Reeve, 1867, Pl. 25, fig. 133.

DESCRIPTION: Shell small, thin, compressed, elongate-ovate, inequilateral, colored with alternating concentric bands of white and pink. Posterior end is bluntly truncate and straight; anterior end is rounded and slightly longer than posterior end. Beaks are low; dorsal margins have gentle slope; ventral margin is broadly arcuate. Postero-umbonal fold is weak; ligament is small and situated in a shallow ligamental groove. Sculpture consists of relatively wide spaced fine concentric ridges which become oblique on the middle of the shell. In the right valve the anterior cardinal is small and lamellar, posterior one is large and bifid; in the left valve the anterior cardinal is short and bifid, posterior one is long and lamellar. There are two inequidistant laterals in the right valve, the posterior one is smaller of the two and situated below the distal end of ligamental groove, the anterior one is well developed and somewhat less distant from the cardinals compared with the posterior one; laterals of left valve are absent. An internal rib extends from below the beak in antero-ventral direction touching posterior margin of anterior adductor muscle scar. Pallial sinus is large, its dorsal margin is posteriorly high and anteriorly descending, its anterior end is sub-rounded and separated from anterior adductor muscle scar only by the internal rib, its ventral margin is entirely coalescent with pallial line. Anterior adductor muscle scar is slightly longer than the posterior one and elongate in shape; posterior one is suborbicular.

FIGURED SPECIMEN: USNM 178517 (Canaries); length, 14.7 mm (100); height, 8.5 mm (57.8); thickness, 3.3 mm (22.4).

HABITAT: Entire coasts of Mediterranean Sea, northwest coast of Africa and Canary Islands.

Subgenus Fabulina Gray

Fabulina Gray, 1851, p. 40.

TYPE: *Tellina fabula* Gronovius (by subsequent designation Winkworth, 1932).

GEOLOGIC RANGE: Recent.

Tellina (Fabulina) fabula Gmelin
(Pl. 9, figs. 1–5)

Tellina fabula Gmelin, 1791, p. 3239; Lamarck, 1818, p. 526; Römer, 1872, p. 132, Pl. 3, figs. 9–11, Pl. 29, figs. 11–14.

DESCRIPTION: Shell small, thin, white, compressed, inequivalve, inequilateral. Anterior end is slightly shorter than posterior end and broadly rounded; posterior end is rostrate with acuminate extremity, and slightly bent to right. Beaks are small and low; antero-dorsal margin is sloping gently, postero-dorsal margin is somewhat steep; ventral margin is broadly arcuate. Postero-umbonal fold is obsolete, a small and lanceolate ligament is situated in a shallow ligamental groove. Sculpture is discrepant; consisting of fine ridges which are oblique on the right valve and concentric on the left valve. Hinge plate is very thin; right posterior and left anterior cardinals are trigonal, bifid, and larger than the other cardinals; right anterior and left posterior cardinals are small and lamellar. There are two inequidistant laterals in right valve; the anterior one is larger than the other one and adjacent to cardinals, posterior one small and situated below the distal end of ligamental groove; laterals of left valve are absent. Pallial sinus is large, its dorsal margin is posteriorly high and anteriorly descending, its anterior end is rounded and extends forward to within one millimeter of anterior adductor muscle scar, its ventral margin is entirely coalescent with pallial line.

FIGURED SPECIMEN: USNM 178196 (Pombrey Burrows, England); length, 19.0 mm (100); height, 11.0 mm (58.0); thickness, 4.0 mm (21.0).

HABITAT: Coast of Great Britain, west coast of Europe from Norway to Portugal, northwest coast of Africa, and Mediterranean.

Subgenus Homalina Stoliczka

Homalina Stoliczka, 1870, p. 118.

Homala Morch, 1852, p. 11 (not *Homala* Agassiz, 1846, nor *Omala* Schumacher, 1817).

TYPE: *Tellina triangularis* Chemnitz, 1782 (=*Tellina trilatera* Gmelin, by original designation).

GEOLOGIC RANGE: Recent.

Tellina (Homalina) trilatera Gmelin
(Pl. 9, figs. 6–10)

Tellina triangularis Chemnitz, 1782, p. 96, Pl. 10, fig. 85 (non-binomid).

Tellina trilatera Gmelin, 1791, p. 3234.

Tellina deltoidalis, var. *b.* Lamarck, 1818, p. 532.

Tellina polita Sowerby, 1825, p. 4.

Tellina triangularis Römer, 1872, p. 179, pl. 3, fig. 1, Pl. 36, figs. 4–6.

DESCRIPTION: Shell small to medium size, white, rather thin, subtrigonal-ovate, compressed, inequilateral. Anterior end is slightly shorter than posterior end and broadly rounded; posterior end is narrow and sharply rounded. Dorsal margins have steep slope, postero-dorsal margin is slightly excavated behind the beaks; ventral margin is broadly arcuate. Ligament is small and situated in a depressed ligamental groove; postero-umbonal fold is obsolete. Sculpture consists of fine concentric ridges which become oblique on the posterior area of right valve. Hinge plate is thin; right valve has a small and sharply pointed anterior cardinal, a bifid posterior cardinal slightly larger than the anterior one and with a deep socket on each side; right anterior lateral is well developed and adjacent to the cardinals, right posterior lateral is somewhat smaller than anterior one and situated below the distal end of ligamental groove. In the left valve the anterior cardinal is bifid, and with a deep socket on each side; the posterior cardinal is small, thin, and partly fused with the nymph; laterals of left valve are obsolete. Pallial sinus is large and discrepant in two valves; in the right valve its dorsal margin is steeply descending anteriorly until it touches the pallial line three millimeters behind the anterior adductor muscle scar and its ventral margin is entirely coalescent with pallial line. In the left valve the dorsal margin of pallial sinus is descending gently and its anterior end is coalescent with the lower margin of anterior adductor muscle scar, its ventral margin is entirely coalescent with pallial line. Anterior adductor muscle scar is slightly larger than posterior one, elongate in shape; posterior one is orbicular.

FIGURED SPECIMEN: USNM 172657; length, 27.2 mm (100); height, 19.5 mm (71.7); thickness, 7.0 mm (25.8).

HABITAT: Pacific and Indian Oceans.

Subgenus *Pharaonella* Lamy

Pharaonella Lamy, 1918, p. 31.

TYPE: *Tellina pharaonis* Hanley (by subsequent designation Salisbury, 1934).

GEOLOGIC RANGE: Recent.

Tellina (Pharaonella) pharaonis Hanley
(Pl. 10, figs. 1–4)

Tellina pharaonis Hanley, 1844, p. 148; Hanley, 1846, p. 235, Pl. 63, fig. 215; Reeve, 1867, Pl. 36, fig. 205; Römer, 1872, p. 61, Pl. 16, figs. 7–10.

DESCRIPTION: Shell of medium size, pink, elongate, narrow, compressed, inequivalve, subequilateral. Anterior end is slightly shorter than posterior end and sharply rounded; posterior end is strongly rostrate, slightly twisted to right, obliquely truncate with acuminate extremity. Beaks are low; antero-dorsal margin is straight and horizontal, postero-dorsal margin is straight and has a gentle slope; ventral margin is broadly arcuate and with a strong sinus at the posterior part. Ligament is short, narrow and situated in a depressed ligamental groove; postero-umbonal fold is weak. Sculpture consists of fine concentric ridges and very fine radial striae. Hinge plate is thin; right anterior and left posterior cardinals are trigonal, bifid, and larger than the other cardinals. Right valve has two equidistant and well developed laterals, the posterior one is situated below the distal end of ligamental groove; in the left valve the posterior lateral is obsolete and anterior one is absent. Pallial sinus is of medium size, about half of it extends beyond middle of the shell, its dorsal margin is low, its anterior end is acutely rounded and 10 millimeters from anterior adductor muscle scar, and 0.87 of its ventral margin is coalescent with pallial line. Adductor muscle scars are subequal and suborbicular.

FIGURED SPECIMEN: USNM 17840 (Ceylon); length, 77.2 mm (100); height, 24.5 mm (31.8); thickness, 9.2 mm (11.9).

HABITAT: Red Sea and coasts of southern India and Ceylon.

Dallitellina, New Subgenus

DIAGNOSIS: Shell of small to medium size, elongate-subtrigonal, narrow, white with some pink coloring, subequilateral, inequivalve. Anterior end is slightly shorter than posterior end and sharply rounded; posterior end is rostrate, obliquely truncate, and bent to right. Dorsal margins have equal and moderately steep slope; ventral margin is broadly arcuate. Ligament is small and situated in a depressed ligamental groove; postero-umbonal fold is sharply defined on both valves. Sculpture consists of fine concentric ridges which terminate at spinous rows along the dorsal margins. There are two cardinals in each valve, right posterior and left anterior cardinals are bifid: right valve has two well developed and distant laterals, laterals of left valve are obsolete. Pallial sinus is small and half of its ventral margin is free.

TYPE: *Tellina rostrata* Linnaeus.

GEOLOGIC RANGE: Recent.

Tellina (Dallitellina) rostrata Linnaeus
(Pl. 10, figs. 5–9)

Tellina rostrata Linnaeus, 1758, p. 675; Hanley, 1847, p. 222, Pl. 61, fig. 157; Reeve, 1866, Pl. 17, fig. 83; Römer, 1871, p. 33, Pl. 3, figs. 5–7, Pl. 11, figs. 4, 5.

Tellina spengleri Lamarck, 1818, p. 522.

DESCRIPTION: Shell of small to medium size, solid, white with some pink coloring, elongate-subtrigonal, narrow, subequilateral, inequivalve. Anterior end is slightly shorter than posterior end and sharply rounded; posterior end is rostrate, bent to right, and obliquely truncate with acuminate extremity. Dorsal margins have equal and moderately steep slope, ventral margin is broadly arcuate. Postero-umbonal fold is sharply defined on both valves; ligament is small, lanceolate, and situated in a depressed ligamental groove; lunule is very small and narrow. Sculpture consists of fine, sharply defined, concentric ridges some of which terminate at an angle on the posterior part of ventral margin; a short space from dorsal margins a row of spines are present on each valve and most of the concentric ridges terminate at these spines, the areas between these rows of spines and dorsal margins are sculptured by fine striae radiating from the beaks. Hinge plate is thin; right anterior and left posterior cardinals are small and lamellar, right posterior and left anterior cardinals are trigonal, bifid, and larger than the other cardinals. Right valve has two well developed laterals, the posterior one is situated below distal end of ligamental groove, anterior one is slightly less distant from the cardinals; laterals of left valve are obsolete. Pallial sinus is small, its dorsal margin is nearly straight, its anterior end is rounded and 18 millimeters from anterior adductor muscle scar, posterior half of its ventral margin is coalescent with pallial line; there is a thin line which joins anterior end of pallial sinus with anterior adductor muscle scar. Anterior adductor muscle scar is slightly larger than posterior one and elongate in shape; posterior one is suborbicular.

FIGURED SPECIMEN: USNM 124695 (Singapore); length, 67.1 mm (100); height, 22.4 mm (33.4); thickness, 11.5 mm (17.1).

HABITAT: Straits of Malacca to Philippine Islands.

Subgenus *Smithsonella*, New Subgenus

DIAGNOSIS: Shell of medium size, elongate-ovate, subequilateral, inequivalve, rose with white rays. Anterior end is slightly shorter than posterior end and sharply rounded; posterior end is obliquely truncate and slightly bent to right. Dorsal margins have equal and gentle slope, ventral margin is broadly arcuate. Postero-umbonal fold is obsolete, a medium size lanceolate ligament is situated in a depressed ligamental groove, lunule is narrow and short. Sculpture consists of fine concentric ridges on the middle of shell and small scales on both ends. There are two cardinals in each valve, right posterior and left anterior cardinals are bifid. Right valve has two well developed and distant laterals, those of left valve are obsolete. Pallial sinus is of medium size and about a third of its ventral margin is coalescent with pallial line. Adductor muscle scars are rather large.

TYPE: *Tellina pulcherrima* Sowerby.

GEOLOGIC RANGE: Recent.

Tellina (Smithsonella) pulcherrima Sowerby
(Pl. 11, figs. 1–5)

Tellina pulcherrima Sowerby, 1825, p. 3, Pl. 1, fig. 1: Hanley, 1847, p. 226, Pl. 61, fig. 165; Reeve, 1867, Pl. 21, fig. 108.

DESCRIPTION: Shell of medium size, elongate-ovate, rose with white rays, subequilateral, inequivalve, left valve more convex than the right. Anterior end is slightly shorter than posterior end and sharply rounded; posterior end is narrower than anterior end, obliquely truncate, and slightly bent to right. Beaks are low; dorsal margins have equal and gentle slope; ventral margin is broadly arcuate and its posterior part is slightly sinuous. Postero-umbonal fold is obsolete; a medium size, lanceolate ligament is situated in a depressed ligamental groove; lunule is narrow and short. Sculpture consists of fine concentric ridges on the middle of the shell, scales of small size on both ends and along ventral margin of right valve. In the right valve the posterior cardinal is large, trigonal, and deeply bifid; anterior cardinal is small, and thin. There are two well developed laterals in right valve; the posterior one is situated below the distal end of ligamental groove, anterior one is slightly less distant from cardinals. In the left valve the anterior cardinal is bifid, well developed but somewhat small; the posterior cardinal is small, thin, and mostly fused with the nymph; laterals of left valve are obsolete. Pallial sinus is of medium size, its dorsal margin is slightly descending anteriorly, its anterior end is acutely rounded and six millimeters from anterior adductor muscle scar, posterior third of its ventral margin is coalescent with pallial line. Adductor muscle scars are subequal, elongate, and large.

FIGURED SPECIMEN: USNM 344678 (Kii, Japan); length, 45.7 mm (100); height, 23.3 mm (51.0); thickness, 10.6 mm (23.2).

HABITAT: Singapore to southern Japan.

REMARKS: Hanley considers two varieties of this species: *pulcherrima*, var. *a.*, which is the one described above, is more elongate, with the concentric ridges and scales stronger, and more numerous. *T. pulcherrima*, var. *b.*, is characterized by having fine and somewhat oblique concentric ridges; scales are present only along the ventral margin.

GENUS QUADRANS BERTIN, 1878

DIAGNOSIS: Shell of medium to moderately large size, white, ovate to subtrigonal-ovate, subequilateral. Anterior end is rounded, posterior end is narrower than anterior end and bluntly truncate. Dorsal margins are generally steep; ventral margin is broadly arcuate; escutcheon is prominent and in most cases bordered by spines, lunule is present in many species. Sculpture is concentric; but in some forms, in addition to concentric, there may be radial or oblique striae. There are two cardinals in each valve; right laterals are typically well developed, laterals of left valve are obsolete. Pallial sinus is generally large.

TYPE: *Tellina gargadia* Linneaus (by subsequent designation Dall, 1900).

Key to The Subgenera of Genus *Quadrans*

A. Sculpture concentric, with or without radial striae.
    a. Pallial sinus coalescent with anterior adductor scar.
        b. Sculpture of concentric ridges and radial striae. . . . . . *Quadrans* s.s.
        bb. Sculpture concentric only.
            (1) Both valves with lateral teeth . . . . *Pristipagia*
            (2) Left valve without lateral teeth . . . . *Acorylus*
    aa. Pallial sinus not coalescent with anterior adductor scar.
        c. Sculpture concentric and radial . . . . *Pistris*
        cc. Sculpture concentric only.
            d. Sculpture of coarse wavy concentric ridges.. . . . . *Quidnipagus*
            dd. Sculpture of fine regular concentric ridges.
                e. Postero-dorsal margin spinous.. . . . . *Phyllodina*
                ee. Postero-dorsal margin not spinous. . . . . *Serratina*
B. Sculpture oblique. . . . . *Obtellina*
    NOTE: The subgenus *Acorylus* Olsson and Harbison, discussed in the appendix, belongs in this genus.

Subgenus *Quadrans* Bertin

Quadrans Bertin, 1878, pp. 265, 266.
TYPE: *Tellina gargadia* Linnaeus (by subsequent designation Dall, 1900).
GEOLOGIC RANGE: Recent.

Quadrans (*Quadrans*) *gargadia* (Linnaeus)
(Pl. 11, figs. 6–10)

Tellina gargadia Linneaus, 1758, p. 674; Lamarck, 1818, p. 530; Hanley, 1846, p. 263, Pl. 61, fig. 156, Pl. 62, fig. 181; Reeve, 1866, Pl. 17, figs. 84a, b; Römer, 1871, p. 38, Pl. 2, figs. 2–4, Pl. 11, figs. 8–10.

DESCRIPTION: Shell of medium size, rather thick, white, subtrigonal-ovate, compressed, inequilateral. Anterior end is slightly longer than posterior end and rounded, posterior end is bluntly truncate and slightly bent to right. Dorsal margins are steeply sloping; ventral margin is convex; postero-umbonal fold is obsolete. Sculpture, on the anterior half of the shell consists of fine concentric ridges which are crossed by widely spaced, very fine radial striae; on the posterior half of the shell the concentric ridges become coarse and wavy, ending in a row of spines on the postero-dorsal margin. There is a small lunule and a prominent escutcheon; the small lanceolate ligament is placed within the upper portion of the escutcheon. The hinge plate is rather thick; right posterior and left anterior cardinals are large and bifid; right anterior and left posterior cardinals are small and lamellar. Right valve has two subequidistant and strong laterals, the posterior one is situated below the distal end of escutcheon, the anterior one is slightly closer to the cardinals; laterals of left valve are obsolete. Pallial sinus is large, its dorsal margin is straight, its anterior end is broadly rounded and coalescent with anterior adductor muscle scar, its ventral margin entirely confluent with pallial line. Adductor muscle scars are subequal, the anterior one is elongate and posterior one is orbicular.

FIGURED SPECIMEN: ANSP 51843 (Java); length, 36.6 mm (100); height, 28.7 mm (78.4); thickness, 9.5 mm (12.1).

HABITAT: Indian Ocean and western Pacific: reported from Zanzibar, Java, Philippines, Japan and New Caledonia.

Subgenus *Pistris* Thiele

Pristis 'Jousseum' Lamy, 1918, p. 29 (not Link 1790, Latham 1794, Müller and Henle 1837).
Pistris Thiele, 1934, p. 917.
TYPE: *Tellina pristis* Lamarck (by tautonymy).
GEOLOGIC RANGE: Recent.

Quadrans (*Pistris*) *pristis* (Lamarck)
(Pl. 12, figs. 1–5)

Tellina pristis Lamarck, 1818, p. 531; Hanley, 1846, p. 268, Pl. 61, fig. 160; Reeve, 1867, Pl. 33, fig. 185 Römer, 1871, p. 41, Pl. 12, figs. 5–7.
DESCRIPTION: Shell of medium size, white, trigonal-ovate, somewhat compressed, equivalve, subequilateral. Anterior end is slightly longer than posterior end and broadly rounded; posterior end is narrow, straight, and bluntly truncate. Dorsal margins are straight and have steep slope, postero-dorsal margin is slightly more steep than the antero-dorsal margin; ventral margin is convex. Postero-umbonal fold is obsolete; escutcheon is large and a medium size lanceolate ligament is situated within the upper third of the escutcheon; lunule is small and mostly on the right valve. Sculpture consists of fine concentric ridges and radial striae; on the posterior third of the shell the concentric ridges become coarse and wavy, the radial striae tend to be weak. There are two cardinals in each valve: right posterior and left anterior cardinals are small and lamellar. Right valve has two subequidistant and well developed laterals, posterior one is situated below the distal end of ligamental groove and anterior one is slightly less distant from the cardinals; laterals of left valve are obsolete. Pallial sinus is large, its dorsal margin posteriorly high and anteriorly descending, its anterior end is rounded and extends forward to within two millimeters of anterior adductor muscle scar, posterior third of its ventral margin is coalescent with pallial line. Anterior adductor muscle scar is larger than posterior one and elongate in shape, posterior one is orbicular.
FIGURED SPECIMEN: USNM 75664 (Philippines); length, 46.4 mm (100); height, 34.5 mm (74.7); thickness, 14.0 mm (30.2).
HABITAT: Indian Ocean and western Pacific: reported from South Africa, Madagascar, Philippines and Australia.

Subgenus *Serratina* Pallary

Serratina Pallary, 1920, p. 95.
Striotellina Thiele, 1934, p. 917.
TYPE: *Tellina serrata* 'Renieri' Brocchi (by original designation).
GEOLOGIC RANGE: Tertiary to Recent.

Quadrans (*Serratina*) *serrata* (Brocchi)
(Pl. 12, figs. 6–10)

Tellina serrata Renieri, 1804, Vol. 6 (nom. nud.); Brocchi, 1814, p. 510, Pl. 12, fig. 1; Hanley, 1846, p. 234, Pl. 66, fig. 257; Reeve, 1868, Pl. 46, fig. 270.
Quadrans serratus Thiele, 1934, p. 917.
DESCRIPTION: Shell of small size, white with faint orange color on the umbonal area, subtrigonal-ovate, slightly compressed, inequivalve, equilateral. Anterior end is broadly rounded; posterior end is narrow, bluntly truncate and slightly bent to right. Dorsal margins have steep slope, postero-dorsal margin is steeper than antero-dorsal margin; ventral margin is broadly arcuate and has a weak emargination at the posterior part. Postero-umbonal fold is obsolete, escutcheon is well developed and a small lanceolate

ligament is situated within the upper part of escutcheon. Sculpture consists of fine concentric ridges that become coarse on the postero-dorsal margin. Right posterior and left anterior cardinals are large and bifid, right anterior and left posterior cardinals are small and lamellar. Right valve has two well developed inequidistant laterals, the posterior one is situated below the distal end of ligamental groove, anterior one is less distant from the cardinals; laterals of left valve are obsolete. Pallial sinus is large, its dorsal margin is posteriorly high and anteriorly descending, its anterior end is rounded and extends forward within one millimeter of anterior adductor muscle scar, ventral margin of it is entirely coalescent with pallial line. Adductor muscle scars are subequal, the anterior one is elongate, posterior one is orbicular.

FIGURED SPECIMEN: USNM 178358 (Zara, Yugoslavia); length, 26.0 mm (100); height, 18.0 mm (69.2); thickness, 7.5 mm (28.8).

HABITAT: Mediterranean Sea south to Canaries.

Subgenus *Phyllodina* Dall

Phyllodina Dall, 1900, p. 290.
TYPE: *Tellina squamifera* Deshayes (by original designation).
GEOLOGIC RANGE: Oligocene to Recent.

Quadrans (Phyllodina) squamifera (Deshayes)
(Pl. 13, figs. 1–5)

Tellina squamifera Deshayes, 1855, p. 365; Reeve, 1869, Pl. 55, fig. 325.
DESCRIPTION: Shell small, white, subtrigonal-ovate, slightly compressed, inequivalve, subequilateral. Anterior end is slightly shorter than posterior end and rounded; posterior end is narrow, bluntly truncate and slightly bent to right. Dorsal margins slope steeply with equal angle on both sides; ventral margin is broadly arcuate and is sinuous at the posterior part. Escutcheon and lunule are well developed; a small lanceolate ligament is situated within the upper part of escutcheon; postero-umbonal fold is obsolete. Sculpture consists of fine concentric ridges which become coarse on the posterior area and terminate at the spines along the postero-dorsal margin. There are two cardinals in each valve; right anterior and left posterior cardinals are small and lamellar, right posterior and left anterior cardinals are trigonal, bifid, and larger than the other cardinals. Right valve has well developed and distant laterals, the posterior one is situated below the distal end of ligamental groove, anterior one is slightly less distant from the cardinals; laterals of the left valve are obsolete. Pallial sinus is large, its dorsal margin is slightly convex, its anterior end is acutely rounded and extends forward to within three millimeters of anterior adductor muscle scar, posterior third of its ventral margin is coalescent with pallial line. Adductor muscle scars are subequal, anterior one is oval, the posterior one orbicular.

FIGURED SPECIMEN: USNM 92196 (Cape Hatteras, North Carolina); length, 22.8 mm (100); height, 14.0 mm (61.4); thickness, 5.4 mm (23.7).

HABITAT: East coast U. S. from Cape Hatteras to Florida and West Indies.

Subgenus *Quidnipagus* Iredale

Quidnipagus Iredale, 1929, p. 266.
TYPE: *Cochlea palatam* Martyn (= *Q. palatam* Iredale) (by original designation).
GEOLOGIC RANGE: Pleistocene to Recent (fossils have been reported from Pleistocene deposits of Red Sea coast and Hawaiian Islands).

Quadrans (*Quidnipagus*) *palatam* (Iredale)
(Pl. 13, figs. 6–10)

Tellina rugosa Born, 1778, p. 29, Pl. 2, figs. 3, 4 (not *Tellina rugosa* Pennant, 1777);
Bryan, 1915, p. 145, Pl. 104, fig. 14.
Cochlea palatam Martyn, 1787, Univ. Conch., Pl. 138 (non-binomid).
Quidnipagus palatam Iredale, 1929, p. 266.

DESCRIPTION: Shell of medium to large size, white, rather thick, trigonal-orbicular, inflated, inequivalve, equilateral. Anterior end is broadly rounded; posterior end is bluntly truncate, and slightly bent to right. The beak of right valve stands slightly higher than that of left valve; dorsal margins have equal and moderately steep slope; ventral margin is convex and slightly sinuous at the posterior part. There is a deep and well developed escutcheon which is occupied by a lanceolate ligament, the lunule is rather short and narrow; postero-umbonal fold is prominent on the right valve and obsolete on the left valve. Sculpture consists of lamellar concentric wavy ridges which become coarse as they approach the margins. Right anterior and left posterior cardinals are small and lamellar, right posterior and left anterior cardinals are trigonal, bifid, and larger than the other cardinals. Right valve has two large and inequidistant laterals each one separated from the margin by a deep furrow, the posterior one is situated below the distal end of the ligamental groove, anterior one is slightly less distant from the cardinals; laterals of left valve obsolete. Pallial sinus is large, its dorsal margin is posteriorly high and anteriorly descending, its anterior end is rounded and extends forward to within one millimeter of anterior adductor muscle scar, posterior half of its ventral margin is coalescent with pallial line. Anterior adductor muscle scar is slightly larger than the posterior one and is elongate in shape, the posterior one is suborbicular.

FIGURED SPECIMEN: USNM 337336 (Honolulu, Hawaii); length, 48.2 mm (100); height, 39.1 mm (81.0); thickness, 19.2 mm (39.9).

HABITAT: Indo-Pacific, from Red Sea and Zanzibar to Hawaiian Islands and Tuamotus.

Subgenus *Pristipagia* Iredale

Pristipagia Iredale, 1936, p. 281.
TYPE: *Pristipagia gemonia* Iredale (by original designation).
GEOLOGIC RANGE: Recent.

Quadrans (*Pristipagia*) *gemonia* (Iredale)
(Pl. 14, figs. 1–2)

Pristipagia gemonia Iredale, 1936, p. 281, Pl. 21, fig. 6.
DESCRIPTION: I was unable to obtain a specimen of *Pristipagia gemonia*; therefore, I am quoting the original description given by Iredale: "Shell small, thin, glassy, white, closely concentrically lirate, no radial striae, almost equilateral, strongly beaked. The hinge is strong, the cardinals prominent and the laterals large and widely separated, the external ligament small, a little sunken. The pallial sinus from muscle to muscle, subparallel to the dorsal angle of the shell."
"Length, 18.5 mm, height, 15.0 mm"
"Habitat: New South Wales. Type from Sydney Harbour."
FIGURED SPECIMEN: Copied from original reference.

Subgenus *Obtellina* Iredale

Obtellina Iredale, 1929, p. 266.

TYPE: *Tellina bougei* Sowerby (by original designation).
GEOLOGIC RANGE: Recent.

Quadrans (*Obtellina*) *bougei* (Sowerby)
(Pl. 14, fig. 3)

Tellina bougei Sowerby, 1909, p. 200, figure.

DESCRIPTION: The specimen of *Tellina bougei* was not available; therefore, I am quoting Sowerby's original description:

"An obliquely oval white shell, with prominent and rather acute umbones, situated rather near the posterior end. The surface of both valves is sculptured with oblique striae, mostly very fine, but becoming coarser near the ventral margin; the sloping posterior dorsal margin is armed with very short blunt spines or scales."

"I know of no species to which this shell bears any close resemblance, though it might be placed near *T. gargadia*, Linn., which is the type of Bertin's section *Quadrans*."

"Hab. - I. Monac, New Caledonia."

FIGURED SPECIMEN: Copied from original reference.

GENUS ARCOPAGIA T. BROWN 1827

DIAGNOSIS: Shell large to small size, orbicular, rounded at the ends, moderately inflated. Posterior flexure is obsolete, beaks are high, and ligament is of medium size. The sculpture is generally concentric, but it may consist of scales or of radial ridges decussating with concentric ones. There are two cardinals in each valve, right posterior and left anterior cardinals are bifid; right valve has two strong laterals, the laterals of left valve are obsolete. Pallial sinus is typically free and ascending, but it may be discrepant in the two valves.

TYPE: *Tellina crassa* Pennant (by subsequent designation Herrmannsen, 1846).
GEOLOGIC RANGE: Eocene to Recent.

Key to the Subgenera of Genus *Arcopagia*

A. Pallial sinus same in both valves.
     a. Ventral margin of pallial sinus entirely free. . . . . *Arcopagia* s. s.
   aa. Ventral margin of pallial sinus partly or entirely coalescent with
       pallial line.
       b. Ventral margin of pallial sinus partly coalescent with pallial
         line.
         c. Sculpture concentric.
           d. Anterior end of pallial sinus connected to the anterior
             adductor scar by a line. . . . . *Johnsonella*,
                                      new subgenus
        dd. Anterior end of pallial sinus not connected to the
          anterior adductor scar by a line.
           e. Sculpture of concentric lamellae.
             f. Posterior end bluntly truncate. . . . . *Macaliopsis*
            ff. Posterior end with acuminate extremity. . . . . *Hemimetis*
          ee. Sculpture of fine concentric striae or ridges.
             g. Shell larger than 25 mm.
               h. Posterior end with strong plication. . . . . . *Sinuosipagia*
              hh. Posterior end without plication. . . . . *Arcopaginula*
            gg. Shell smaller than 25 mm.

            i. Shell thick. . . . . *Cyclotellina*
            ii. Shell thin.
                j. Without posterior fold.
                    k. Concentric ridges somewhat lamellar.   . . *Arcopagiopsis*
                    kk. Concentric ridges low. . . . . *Pinguitellina*
                jj. With posterior fold on right valve. . . . . . *Punipagia*
        cc. Sculpture consisting of scales.
            l. Shell orbicular. . . . . *Scutarcopagia*
            ll. Shell elongae-ovate. . . . . *Smitharcopagia*,
                                                                 new subgenus
    bb. Ventral margin of pallial sinus entirely coalescent with
        pallial line.
        m. Shell very thin and hyaline. . . . . *Merisca*
        mm. Shell thick and porcelaneous.
            n. Anterior end of pallial sinus coalescent with the base of
                anterior adductor scar. . . . . *Macomona*
            nn. Anterior end of pallial sinus not coalescent with
                anterior adductor scar.
                o. Sculpture decussate. . . . . *Pseudarcopagia*
                oo. Sculpture concentric.
                    (1) With strongly sculptured rostral area . . . . . *Zearcopagia*
                    (2) Without sculptured rostral area . . . . *Lyratellina*
B. Pallial sinus discrepant in two valves.
    a. Sculpture concentric only. . . . . *Arcopella*
    aa. Sculpture both radial and concentric. . . . . *Clathrotellina*
    NOTE: *Lyratellina* Olsson, discussed in the appendix, is a subgenus *Arcopagia*.

Subgenus *Arcopagia* T. Brown

Arcopagia (Leach MS.) T. Brown, 1827, p. II, Pl. 16.
Cydippe Leach, 1852, p. 314 (not *Cydippe* Eschscholtz, 1829).
TYPE: *Tellina crassa* Pennant (by subsequent designation Herrmannsen 1846).
GEOLOGIC RANGE: Recent.

Arcopagia (*Arcopagia*) *crassa* (Pennant)
(Pl. 14, figs. 4–8)

Tellina crassa Pennant, 1777, p. 87, Pl. 28, fig. 28; Lamarck, 1818, p. 529; Römer, 1871, p. 80, Pl. 22, figs. 7–10.
Tellina rigida Donovan, 1801, Pl. 103.
Tellina maculata Turton, 1819, p. 173, fig. 13.
Arcopagia ovata Brown, 1827, p. 99, Pl. 11, figs. 9–10.

DESCRIPTION: Shell of medium size, white, thick, orbicular, somewhat inflated, inequivalve, inequilateral. Anterior end is longer than the posterior end and rounded; posterior end is rounded and slightly broader than anterior end. Beaks are situated a short distance posterior of mid-length, dorsal margins are moderately steep; ventral margin is broadly arcuate and has a feeble emargination at the posterior part. A medium size, lanceolate ligament is situated in a depressed ligamental groove; the lunule is well defined but small. Sculpture consists of fine concentric ridges which become somewhat irregular on the anterior part of the shell. There are two cardinals in each valve; right anterior and left posterior cardinals are small and thin; right posterior and left anterior cardinals are trigonal, bifid, and larger than other cardinals. Right valve has two inequidistant laterals, the posterior one is less distant from the cardinals and larger than the posterior one; laterals of left valve are obsolete. Pallial sinus is of medium size, obliquely ascending and entirely free. An internal rib extends from beneath the beak to the posterior margin of anterior adductor muscle scar. Posterior adductor muscle scar is slightly larger than anterior one and suborbicular in shape, anterior one is oval.

FIGURED SPECIMEN: USNM 178285 (Mediterranean); length, 48.5 mm (100); height, 40.8 mm (84.4); thickness, 19.6 mm (40.4).

HABITAT: North Sea, Atlantic coast of Europe, and Mediterranean Sea.

Subgenus *Arcopaginula* Lamy

Arcopaginula Lamy, 1918, p. 167.

TYPE: *Tellina inflata* Gmelin (by original designation).

GEOLOGIC RANGE: Recent.

Arcopagia (Arcopaginula) inflata (Gmelin)
(Pl. 15, figs. 1–5)

Tellina inflata Gmelin, 1791, p. 3230; Römer, 1871, p. 52, Pl. 15, figs. 1–3.

Tellina ovata Röding, 1798, p. 2.

Tellina hippoidea Jonas, in Philippi, 1844, p. 72, Pl. 1, fig. 3.

DESCRIPTION: Shell of medium size, white, orbicular, moderately inflated, subequilateral. Anterior end is slightly longer than posterior end and broadly rounded; posterior end is acutely rounded and slightly bent to right. Dorsal margins have moderately steep slope, postero-dorsal margin is more steep than the antero-dorsal margin; ventral margin is convex. A small lanceolate ligament is situated in a depressed ligamental groove, and the lunule is very small. Sculpture consists of fine concentric ridges and some widely spaced, very fine radial striae. Right anterior and left posterior cardinals are small and lamellar, right posterior and left anterior cardinals are trigonal, bifid, and larger than the other cardinals. Right valve has two strong and inequidistant laterals, the posterior one is situated below the distal end of ligamental groove, and anterior one is less distant from cardinals; laterals of left valve are obsolete. Pallial sinus is of medium size, its dorsal margin is posteriorly ascending and anteriorly descending, its anterior end is acutely rounded and eleven millimeters from anterior adductor muscle scar, posterior half of its ventral margin is coalescent with pallial line. Anterior adductor muscle scar is slightly larger than the posterior one and has oval shape, the posterior one is suborbicular.

FIGURED SPECIMEN: USNM 168730 (Australia); length, 46.1 mm (100); height, 36.8 mm (80.0); thickness, 16.9 mm (36.6).

HABITAT: Australia, Philippines and China Seas.

Subgenus *Johnsonella*, New Subgenus

DIAGNOSIS: Shell large, thick, white, orbicular, somewhat inflated, subequilateral. Both ends are rounded, dorsal margins slope down steeply with an equal angle, ventral margin is convex. Beaks are situated slightly posterior to mid-length, ligament is stout and situated in a rather shallow ligamental groove; sculpture consists of fine concentric ridges. There are two cardinals in each valve, right posterior and left anterior cardinals are bifid; laterals of right valve are strong, and those of left valve are obsolete. Pallial sinus is of medium size and half of its ventral margin is coalescent with pallial line.

TYPE: *Tellina fausta* Pultney.

GEOLOGIC RANGE: Recent.

Arcopagia (Johnsonella) fausta (Pultney)
(Pl. 15, figs. 6–7; Pl. 16, figs. 1–3)

Tellina fausta Pultney, 1799, p. 29, Pl. 5, fig. 5; Donovan, 1801, Pl. 98, Hanley, 1846, p. 256, Pl. 64, figs. 230, 234; Römer, 1871, p. 77, Pl. 21, figs. 4–7.

DESCRIPTION: Shell large, thick, white, orbicular, somewhat inflated, subequilateral. Anterior end is slightly longer than posterior end and broadly rounded, posterior end is bluntly truncate. Beaks are prominent, dorsal margins slope down steeply with an equal angle on both sides, ventral margin is convex. A stout ligament is situated in a shallow ligamental groove; the sculpture consists of fine concentric ridges and some very fine, widely spaced striae. Hinge plate is of medium strength, and there are two cardinals in each valve; right anterior and left posterior cardinals are small and lamellar, right posterior and left anterior cardinals are trigonal, bifid, and larger than the other cardinals. Right valve has two strong and inequidistant laterals, the posterior one is situated below the distal end of ligamental groove, anterior one is closer to the cardinals compared with the posterior one; laterals of left valve are obsolete. Pallial sinus is of medium size, its dorsal margin is descending anteriorly, its anterior end is acutely rounded and extends forward to within eleven millimeters of anterior adductor muscle scar and connected to it by a line, the posterior half of its ventral margin is coalescent with pallial line. Anterior adductor muscle scar is larger than posterior one and has elongate shape, posterior one is suborbicular.

FIGURED SPECIMEN: USNM 3206 (St. Thomas, West Indies); length, 72.8 mm (100); height, 63.8 mm (87.0); thickness, 30.2 mm (41.5).

HABITAT: West Indies.

Subgenus *Scutarcopagia* Pilsbry

Scutarcopagia Pilsbry, 1918, p. 332
TYPE: *Tellina scobinata* Linnaeus (by original designation).
GEOLOGIC RANGE: Pleistocene (Red Sea Coast) to Recent.

Arcopagia (Scutarcopagia) scobinata (Linnaeus)
(Pl. 17, figs. 1–5.)

Tellina scobinata Linnaeus, 1758, p. 676; Lamarck, 1818, p. 529; Hanley, 1846, p. 266, Pl. 64, fig. 236; Römer, 1871, p. 73, Pl. 20, figs. 5–8.

DESCRIPTION: Shell of moderately large size, thick, orbicular, inflated, subequilateral, white with some brown discontinuous rays. Anterior end is slightly shorter than posterior end and broadly rounded, posterior end is bluntly truncate. Beaks are prominent and slightly prosogyrate, dorsal margins slope down steeply with an equal angle on both sides, ventral margin is convex. Postero-umbonal fold is weak on right valve and obsolete on the left; ligament is stout and situated in a deep escutcheon, lunule is large and mostly on the left valve. Sculpture consists of strong and erect scales, those on the right valve are slightly larger than those on the left valve. Right anterior and left posterior cardinals are small and lamellar, right posterior and left anterior cardinals are trigonal, bifid, and larger than the other cardinals. Right valve has two strong and subequidistant laterals, the posterior one is situated below the distal end of ligamental groove, anterior one is slightly less distant from the cardinals compared with the posterior one; laterals of left valve are obsolete. Pallial sinus is of medium size, its dorsal margin is descending anteriorly, its anterior end is acutely rounded and extends forward to within eight millimeters of anterior adductor muscle scar and connected to it by a thin line, the posterior third of its ventral margin is coalescent with pallial line.

Anterior adductor muscle scar is larger than the posterior one and has oval shape, posterior one is sub-orbicular.

FIGURED SPECIMEN: USNM 363377 (Tiarei, Tahiti); length 55.3 mm (100); height, 54.5 mm (98.5); thickness, 27.7 mm (50.0).

HABITAT: Indian and Pacific Oceans from Red Sea to Hawaiian Islands and Society Islands.

Subgenus Smitharcopagia, New Subgenus

DIAGNOSIS: Shell of medium size, elongate-ovate, subequilateral, slightly inflated, white with rose coloring on the umbones and some interrupted faint rays of the same color. Anterior end is slightly longer than posterior end and broadly rounded, the posterior end is bluntly truncate. Beaks are low, lunule is of medium size, postero-umbonal fold is weak, and a medium size ligament is situated in a deep escutcheon. Sculpture consists of erect scales which become very small in size as they approach the umbones. There are two cardinals in each valve; right valve has two strong and sub-equidistant laterals, laterals of left valve are obsolete. Pallial sinus is moderately large, its anterior end is connected to the anterior adductor muscle scar by a thin line, half of its ventral margin is coalescent with pallial line.

TYPE: *Tellina linguafelis* Linnaeus.

GEOLOGIC RANGE: Recent

Arcopagia (Smitharcopagia) linguafelis (Linneaus)
(Pl. 18, figs. 1–5)

Tellina linguafelis Linnaeus, 1758, p. 674; Lamarck, 1818, p. 530.

Tellina lingua-felis Hanley, 1846, p. 266, Pl. 64, fig. 236; Römer, 1872, p. 50, Pl. 2, fig. 5, Pl. 14, figs. 7–10.

DESCRIPTION: Shell of medium size, elongate-ovate, subequilateral, slightly inflated, white with rose coloring on the umbones and some interrupted faint rays of the same color. Anterior end is slightly longer than the posterior end and broadly rounded, posterior end is bluntly truncate. Beaks are somewhat low and slightly prosogyrate, antero-dorsal margin has gentle slope, postero-dorsal margin is rather steep, ventral margin broadly arcuate. Ligament of medium size is situated in a deep escutcheon, lunule is moderately large and equally divided in two valves, postero-umbonal fold is weak on the right valve and obsolete on the left. Sculpture consists of erect scales which diminish in size considerably as they approach the umbones. Right anterior and left posterior cardinals are small and lamellar; right posterior and left anterior cardinals are trigonal, bifid, and larger than the other cardinals. Right valve has two strong and inequidistant laterals, the posterior one is situated below the distal end of ligamental groove, anterior one is less distant from the cardinals compared with the posterior one; laterals of left valve are obsolete. Pallail sinus is moderately large, its dorsal margin is descending anteriorly, its anterior end is acutely rounded and connected to the anterior adductor muscle scar by a thin line, posterior half of its ventral margin is coalescent with pallial line. Anterior adductor muscle scar is slightly larger than posterior one and has oval shape, the posterior one is suborbicular.

FIGURED SPECIMEN: USNM 76525 (Northeast coast of Australia); length, 52.0 mm (100); height, 36.9 mm (71.0); thickness, 16.0 mm (30.8).

HABITAT: Indian and Western Pacific Oceans from Indonesia and northern Australia to Ryukyus and Palau Islands.

Subgenus Macaliopsis Cossmann

Macaliopsis Cossmann, 1886, p. 75.
TYPE: *Tellina barrandei* Deshayes (by subsequent designation Dall, 1900).
GEOLOGIC RANGE: Eocene.

Arcopagia (*Macaliopsis*) barrandei (Deshayes)
(Pl. 18, figs. 6–8)

Tellina barrandei Deshayes, 1860, p. 344, Pl. 27, figs. 18–20; Cossmann, 1886, p. 76.
DESCRIPTION: Shell moderately large, subtrigonal-orbicular, somewhat inflated, equilateral. Anterior end is broadly rounded, posterior end is bluntly truncate; antero-dorsal margin has gentle slope, postero-dorsal margin is moderately steep, and ventral margin is broadly convex. Escutcheon is deep and extends along the entire length of postero-dorsal margin, lunule is of medium size and well defined; postero-umbonal fold is strong. Sculpture consists of concentric thin lamellae with interspaces between them being slightly wider than the thickness of the lamellae. There are two cardinals in each valve, right posterior and left anterior cardinals are bifid; right valve has two sub-equidistant and well developed laterals, the posterior one is situated below the distal end of escutcheon, and anterior one is slightly less distant from the cardinals compared to the posterior one; laterals of left valve are obsolete. Pallial sinus is of medium size, somewhat ascending, its anterior end is rounded and extends forward within nine millimeters of the anterior adductor muscle scar, and posterior half of its ventral margin is coalescent with pallial line. Anterior adductor muscle scar is elongate, and posterior one is suborbicular.
FIGURED SPECIMEN: The figures shown here are copied from the original illustrations of Deshayes. The measurements are given as: length, 40.0 mm; height, 32.0 mm (local., Parnes, France).
OCCURRENCE: Calcaire grossier formation, Eocene, France

Subgenus Hemimetis Thiele

Hemimetis Thiele, 1934, p. 915.
TYPE: *Tellina plicata* Valenciennes (= *angulata* Linnaeus) (by monotypy)
GEOLOGIC RANGE: Recent.

Arcopagia (*Hemimetis*) angulata (Linnaeus)
(Pl. 19, figs. 1–5),

Tellina angulata Linnaeus 1767, p. 1116.
Tellina plicata 'Val.' Bory de St. Vincent, 1827, p. 154; Hanley, 1846, p. 270. Pl. 62, fig. 191; Römer, 1871, p. 86, Pl. 23, figs. 7–9.
DESCRIPTION: Shell large, white, solid, somewhat thin, trigonal-orbicular, slightly inflated, equivalve. Anterior end is broadly rounded, posterior end is bluntly truncate, postero-umbonal fold is strong on the right valve and obsolete on the left valve. Antero-dorsal margin has a gentle slope and is convex except immediately in front of the beak where it is excavated, postero-dorsal margin is straight and has moderately steep slope; ventral margin is broadly arcuate and strongly sinuous at the posterior part. Ligament is long, lanceolate and situated in a depressed ligamental groove; lunule is rather small but well defined and most of it is on the right valve; beaks are prosogyrate. Sculpture consists of fine concentric lamellae with interspaces twice as wide as the thickness of lamellae. Right posterior and left anterior cardinals are large, trigonal, and bifid; right anterior cardinal is small, short, and pointed; left posterior

cardinal is long, thin and lamellar. Right valve has two well developed laterals, the posterior one is situated below the distal end of ligamental groove and anterior one is larger than the posterior one and closer to cardinals. Pallial sinus is large and somewhat ascending, its anterior end is broadly rounded and extends forward to within nine millimeters of anterior adductor muscle scar, and posterior half of its ventral margin is coalescent with pallial line. Anterior adductor muscle scar is elongate and the posterior one is suborbicular.

FIGURED SPECIMEN: USNM 538502 (New Caledonia); length, 61.0 mm (100); height, 47.3 mm (77.5); thickness, 21.9 mm. (35.9).

HABITAT: Indian Ocean and Western Pacific; reported from Madagascar, Malacca, New Caledonia and Fiji Islands.

Subgenus *Sinuosipagia* Cossmann

Sinuosipagia Cossmann, 1921, p. 41.

TYPE: *Tellina colpodes* Bayan (by original designation).

GEOLOGIC RANGE: Middle and Upper Eocene.

Arcopagia (Sinuosipagia) colpodes (Bayan)
(Pl. 20, figs. 1–2)

Tellina sinuata Lamarck, 1806, p. 233 (not Spengler, 1795); Lamarck, 1808, Pl. 40, fig. 8a, b; Deshayes, 1824, p. 79, Pl. 11, figs. 15–16; Deshayes, 1832, p. 1018; Deshayes, 1860, p. 348.

Tellina colpodes Bayan, 1873, Pt. 2, p. 119; Cossmann, 1886, p. 77; Cossmann, 1921, p. 41.

Tellina curva—Salisbury, 1934, p. 89.

DESCRIPTION: Shell moderately large, trigonal-orbicular, rather inflated, inequilateral, inequivalve. Anterior end is long and broadly rounded; posterior end is bluntly truncate, plicate, and sinuous; right valve is more inflated than left valve. Beaks are high and situated 0.4 of the length from posterior end; antero-dorsal margin has a moderately steep slope, postero-dorsal margin is very steep; ventral margin is broadly arcuate at the anterior part and sinuous at the posterior. Sculpture consists of fine concentric ridges. There are two cardinals in each valve; right posterior and left anterior cardinals are bifid and larger than other cardinals. Right valve has two strong and distant laterals, in the left valve the laterals are obsolete. Pallial sinus is ascending and similar to that of *T. crassa;* adductor muscle scars are inequal.

FIGURED SPECIMEN: The figures of this species shown here are copied from those appearing in Deshayes' 1824 work. The measurements are given as: length, 35 mm, height 27 mm.

OCCURRENCE: Calcaire grossier formation, at Grignon near Paris.

Subgenus *Pseudarcopagia* Bertin

Pseudarcopagia Bertin, 1878, pp. 229, 264.

TYPE: *Tellina decussata* Lamarck (= *Tellina victoriae* Gatliff and Gabriel) (by subsequent designation Dall, 1900).

GEOLOGIC RANGE: Recent.

Arcopagia (Pseudarcopagia) victoriae (Gatliff and Gabriel)
(Pl. 20, figs. 3–7)

Tellina decussata Lamarck 1818, p. 532, (not *Tellina decussata* Wood, 1815); Hanley, 1846, p. 262, Pl. 60, fig. 184; Römer, 1871 p. 83, Pl. 23, figs. 1–3.

Tellina victoriae Gatliff and Gabriel, 1914, p. 83.

DESCRIPTION: Shell of medium size, thick, white, trigonal-orbicular, inflated, equilateral. Anterior and posterior ends are rounded; dorsal margins slope down steeply with an equal angle on both sides, ventral margin is convex. Beaks are prominent, postero-umbonal fold is obsolete, ligament is of medium size and situated in a depressed ligamental groove, lunule is small and mostly on the right valve. Sculpture is decussate, consisting of fine concentric ridges and radial striae. Hinge plate is thick; right anterior and left posterior cardinals are small and thin, right posterior and left anterior cardinals are bifid and larger than the other cardinals. Right valve has two strong laterals, the posterior one is closer to the cardinals compared to the anterior one; laterals of left valve are obsolete. Pallial sinus is large, its dorsal margin is posteriorly high and anteriorly descending until it touches the pallial line about ten millimeters behind the anterior adductor muscle scar, the ventral margin of it is entirely coalescent with pallial line. Adductor muscle scars are subequal, the anterior one is oval and posterior one is suborbicular.

FIGURED SPECIMEN: USNM 160232 (South Australia); length, 47.1 mm (100); height, 45.2 mm (96.0); thickness, 24.0 mm (50.9).

HABITAT: Western Australia, South Australia, and Tasmania.

Subgenus *Zearcopagia* Finlay

Zearcopagia Finlay, 1926, p. 266.

TYPE: *Tellina disculus* Deshayes (by original designation).

GEOLOGIC RANGE: Pliocene to Recent.

Arcopagia (*Zearcopagia*) *disculus* (Deshayes)
(Pl. 21, figs. 1–5)

Tellina disculus Deshayes, 1854, p. 360; Reeve, 1868, Pl. 52, fig. 306; Römer, 1871, p. 79, Pl. 22, figs. 4–6.

DESCRIPTION: Shell of medium size rather thick, trigonal-orbicular, inflated, white with yellow coloring on the beaks. Anterior end is broadly rounded, posterior end is bluntly truncate; dorsal margins slope down steeply with equal angle on both sides, ventral margin is convex. The ligament is of medium size and situated in a depressed ligamental groove, postero-umbonal fold is obsolete. Sculpture consists of fine concentric ridges which become slightly coarse near the margins. Right anterior and left posterior cardinals are small and thin, right posterior and left anterior cardinals are trigonal, bifid and larger than the other cardinals. Right valve has two strong laterals, the posterior one is situated below the distal end of ligamental groove, anterior one is slightly larger than the posterior one and closer to the cardinals compared with the other lateral; laterals of left valve are obsolete. The interior is colored yellow; pallial sinus is large, its dorsal margin is posteriorly high and anteriorly descending until it touches pallial line about five millimeters behind the anterior adductor muscle scar, its ventral margin is entirely coalescent with pallial line. Adductor muscle scars are subequal, the anterior one is elongate, and posterior one is suborbicular.

FIGURED SPECIMEN: USNM 593282 (Takapuna, Auckland, N. Z.); length, 42.6 mm (100); height, 37.5 mm (88.0); thickness, 17.5 mm (41.2).

HABITAT: New Zealand.

Subgenus *Macomona* Finlay

Macomona Finlay, 1926, p. 466.

TYPE: *Tellina liliana* Iredale (by original designation).
GEOLOGIC RANGE: Recent.

Arcopagia (Macomona) liliana (Iredale)
(Pl. 21, figs. 6–10)

Tellina lactea Quoy and Gaimard, 1815, p. 501, Pl. 81, figs. 14–16 (not *Tellina lactea* Linnaeus).
Tellina liliana Iredale, 1915, p. 488.

DESCRIPTION: Shell moderately large, solid, white with some faint yellow coloring, trigonal-ovate, rather compressed, inequilateral, inequivalve with right valve more convex than the left. Anterior end is broadly rounded; posterior end is narrower than the anterior end, obliquely truncate and slightly bent to right. Beaks are opisthogyrate and situated slightly anterior of the mid-length; antero-dorsal margin is convex and has gentle slope, postero-dorsal margin has moderately steep slope and is convex behind the beaks; ventral margin is broadly arcuate and slightly sinuous at the posterior part. Postero-umbonal fold is prominent on the right valve and weak on the left valve; ligament is of medium size and situated in a depressed ligamental groove. The sculpture consists of fine concentric striae. Right anterior and left posterior cardinals are small and thin; right posterior and left anterior cardinals are trigonal, bifid, and larger than the other cardinals. Right valve has two laterals; the posterior one is small and situated below the distal end of the ligamental groove, anterior one is well developed and is closer to the cardinals than is the posterior one; laterals of left valve are obsolete. Pallial sinus is large; its dorsal margin is posteriorly high and anteriorly descending until it becomes coalescent with the lower end of anterior adductor muscle scar, and ventral margin of it is entirely confluent with pallial line. Anterior adductor muscle scar is very elongate in shape; posterior one is suborbicular.

FIGURED SPECIMEN: USNM 22833 (New Zealand); length, 56.0 mm (100); height, 45.4 mm (81.2); thickness, 17.3 mm (30.9).

HABITAT: New Zealand.

REMARKS: *Tellina liliana* Iredale, a New Zealand species, and *Tellina deltoidalis* Lamarck 1818, are two closely related species. Lamarck in his 1818 work, Anim. S. Vert., vol. 5, pp. 532, 533, considered both as conspecific and included them within the species *Tellina deltoidalis*. Quoy and Gaimard in Voy. Astrol., vol. 3, p. 501, pl. 81, figs. 14–16, described the New Zealand species under the name *Tellina lactea;* because this name is preoccupied by *Tellina lactea* Linnaeus, Syst. Nat. Ed. 10, p. 676, 1758, Iredale in 1915 proposed *Tellina liliana*, a new name for this species. Suter used *T. deltoidalis* for the Australian species, but considered the Australian and New Zealand species to be identical. Now it is generally considered by E. A. Smith, T. Iredale, H. J. Finlay and others that the New Zealand species is to be identified under the name *Tellina liliana* Iredale, and the name *Tellina deltoidalis* Lamarck retained for the Australian species.

Subgenus *Cyclotellina* Cossmann

Cyclotellina Cossmann, 1886, p. 79.
TYPE: *Tellina lunulata* Lamarck (by original designation).
GEOLOGIC RANGE: Eocene.

Arcopagia (Cyclotellina) lunulata (Lamarck)
(Pl. 22, figs. 1–4)

Donax lunulata Lamarck, 1805, p. 139; Lamarck, 1808, Pl. 41, figs. 5a, b.

Tellina lunulata Deshayes, 1824, p. 79, Pl. 11, figs. 3, 4; Deshayes, 1832, p. 1018; Deshayes, 1835, p. 212; d'Orbigny, 1850, p. 377; Deshayes, 1860, p. 354.

DESCRIPTION: Shell small, orbicular, subequilateral, slightly compressed. Anterior end is short and broadly rounded, posterior end is rounded and narrower than the anterior end; dorsal margins slope down steeply, the postero-dorsal margin is slightly more steep than the antero-dorsal margin; ventral margin is broadly arcuate. Beaks are situated slightly anterior of mid-length; ligamental groove is lanceolate in shape and small. Sculpture consists of fine concentric striae. Right anterior and left posterior cardinals are small and partly fused with the nymph, right posterior and left anterior cardinals are trigonal, bifid and larger than the other cardinals. Right valve has two well developed laterals, the posterior one is situated below the distal end of ligamental groove, anterior one is slightly larger than the posterior one and closer to the cardinals; laterals of left valve are obsolete. Pallial sinus is large, its dorsal margin descending anteriorly, its anterior end is rounded and extends forward to within one millimeter of anterior adductor muscle scar, and posterior half of its ventral margin is coalescent with pallial line. Anterior adductor muscle scar is slightly larger than posterior one and has elongate shape, posterior one is suborbicular.

FIGURED SPECIMEN: USNM 142300 (Anvers, Paris Basin); length, 19.8 mm (100); height, 18.4 mm (93.0); thickness, 8.0 mm (40.4).

OCCURRENCE: Eocene, Paris Basin and Belgium.

Subgenus *Arcopagiopsis* Cossmann

Arcopagiopsis Cossmann, 1886, p. 81.

TYPE: *Tellina pustula* Deshayes (by subsequent designation Dall, 1900).

GEOLOGIC RANGE: Lower Eocene.

Arcopagia (Arcopagiopsis) pustula (Deshayes)
(Pl. 22, figs. 5–9)

Tellina pustula Deshayes, 1824, p. 85, Pl. 13, figs. 9–11; Deshayes, 1832, p. 1020; Bronn, 1848, p. 1222; Deshayes, 1860, p. 356.

Arcopagia pustula d'Orbigny, 1848, p. 276.

DESCRIPTION: Shell small, thin, white, orbicular, somewhat inflated, subequilateral, anterior and posterior ends rounded. Beaks are rather high and situated slightly posterior of mid-length; dorsal margins have equal and moderately steep slope, the antero-dorsal margin is slightly excavated in front of the beak; ventral margin is convex. Ligamental groove is of medium size, lunule is rather large, and postero-umbonal fold is absent. Sculpture consists of somewhat widely spaced, thin, lamellae-like, concentric ridges with interspaces between them about four times as wide as the thickness of the ridges. In the right valve the anterior cardinal is thin, sharply pointed and with a deep socket on each side; posterior cardinal is trigonal, strongly bifid, with a deep socket on each side, and about twice as large as the anterior one. Laterals of right valve are strong and each one is separated from the shell margin by a deep furrow, the posterior one is situated below the distal end of ligamental groove and anterior one is stronger than the posterior one and less distant from cardinals. Adductor muscle scars are subequal and suborbicular.

FIGURED SPECIMEN: USNM 142302 (Mouchy, Oise, Paris Basin); length, 4.0 mm (100); height, 2.7 mm (67.4); thickness, 1.0 mm (25.0).

OCCURRENCE: Calcaire grossier formation, France.

REMARKS: In the collections of this Museum only one right valve of this species is present, but it does not show the characteristics of pallial sinus and in the previous works no mention has been made about it. Because I have not seen the left valve I am unable to present my own observations on that valve. Deshayes mentions that the left valve has only one cardinal tooth; however, from my observations of the left valve of smaller species of this family, I am inclined to think that the left valve of this species has two cardinals, one being obsolete. Two cardinals in each valve is an invariable morphologic character of all species of the Tellinidae, but in small species, whenever the right posterior cardinal is unusually large, which is the case in this species, the socket for that tooth on the left valve becomes so large that the left posterior cardinal is crowded toward the nymph in such a way that it becomes obsolete and partly fused to the nymph.

Subgenus *Arcopella* Thiele

Arcopella Thiele, 1934, p. 914.
TYPE: *Tellina balaustina* Linnaeus (by monotypy).
GEOLOGIC RANGE: Recent.

Arcopagia (Arcopella) balaustina (Linnaeus)
(Pl. 22, figs. 10–14)

Tellina balaustina Linnaeus, 1758, p. 676; Hanley, 1846, p. 253, Pl. 56, fig. 10; Römer, 1871, p. 92, Pl. 24, figs. 10–12.
Lucina balaustina Payraudeau, 1827, p. 43, Pl. 1, figs. 21–22.

DESCRIPTION: Shell small, thin, white, orbicular, inflated, equilateral, anterior and posterior ends rounded. Beaks are rather high, dorsal margins have equal slope on both sides and are moderately steep; ventral margin is convex. Ligament is small and situated in a shallow ligamental groove, lunule is small and mostly on the left valve; postero-umbonal fold is obsolete. Sculpture consists of fine concentric ridges with interspaces about twice as wide as the thickness of the ridges. Hinge plate is thin; right posterior cardinal is rather large, trigonal, and bifid; right anterior cardinal is small and lamellar; left anterior cardinal is small and bifid; left posterior cardinal is lamellar and partly fused to the nymph. Right valve has two well developed laterals, the posterior one is smaller than anterior one and situated below the distal end of ligamental groove, anterior one is strong and closer to the cardinals compared to the posterior one; laterals of left valve are obsolete. The pallial sinus is discrepant in the two valves, in right valve it is small, its dorsal margin slightly descending, the anterior end of it is acutely rounded and extends forward only to the middle of the shell, and two-thirds of its ventral margin is free; the pallial sinus of left valve is large, dorsal margin of it is slightly descending, its anterior end is rounded and touches the lower part of anterior adductor muscle scar, and two-thirds of its ventral margin is free. Adductor muscle scars are subequal and suborbicular.

FIGURED SPECIMEN: USNM 178454 (off Ireland); length, 21.8 mm (100); height, 17.1 mm (78.0); thickness, 8.8 mm (25.0).

HABITAT: Northeastern Atlantic, from Ireland to the Canary Islands; in the Mediterranean it is found on the coasts of Sicily, Corsica, Algiers, France and Spain.

Clathrotellina Thiele, 1934, p. 917.

Subgenus *Clathrotellina* Thiele

TYPE: *Merisca pretiosa* (Deshayes) (= *Tellina pretium* Salisbury) (by monotypy).
GEOLOGIC RANGE: Recent.

Arcopagia (Clathrotellina) pretium (Salisbury)
(Pl. 22, figs. 15–19)

Tellina pretiosa Deshayes, 1854, p. 360 (not Eichwald 1830); Reeve, 1869, Pl. 56, fig. 329.

Tellina pretium Salisbury, 1934, p. 86.

DESCRIPTION: Shell small, thin, white, ovate-subtrigonal, inflated, equilateral. Anterior end is rounded, posterior end is acutely rounded; dorsal margins slope down on both sides at a steep and equal angle, ventral margin is convex. Beaks are rather high, ligament is small and situated in a depressed ligamental groove, lunule is small and mostly on the left valve. Sculpture consists of fine concentric ridges, and somewhat widely spaced, prominent, erect radial ridges bearing denticles produced by the inter-section of concentric ridges by radials. Right anterior cardinal is small and lamellar, right posterior cardinal is trigonal, bifid, and larger than the other cardinals; in the left valve the anterior cardinal is rather small and bifid, posterior cardinal is small and partly fused to the nymph. Right valve has two inequidistant, well developed laterals, the posterior one is situated below the distal end of ligamental groove and anterior one is slightly less distant from the cardinals; laterals of left valve are obsolete. The pallial sinus is discrepant in two valves, that of right valve is small, its dorsal margin steeply descending, its anterior end is acutely rounded and extends forward only up to the middle of the shell, posterior half of its ventral margin is coalescent with pallial line; pallial sinus of left valve is large, its dorsal margin is nearly straight, its anterior end is broadly rounded and coalescent with anterior adductor muscle scar, and posterior half of its ventral margin is coalescent with pallial line. Adductor muscle scars are subequal and orbicular.

FIGURED SPECIMEN: USNM 257839 (Tinagta Id., Tawi Tawi Ids., Philippines); length, 16.5 mm (100); height, 13.0 mm (78.8); thickness, 7.0 mm (42.4).

HABITAT: Philippine Islands.

Subgenus *Merisca* Dall

Merisca Dall, 1900, p. 290.

TYPE: *Tellina crystallina* Wood (= *Tellina cristallina* Spengler) (by original designation).

GEOLOGIC RANGE: Miocene to Recent. Fossils are reported from Miocene deposits of Banana River, Costa Rica.

Arcopagia (Merisca) cristallina (Spengler)
(Pl. 23, figs. 1–5)

Tellina cristallina Spengler, 1798, p. 113.

Tellina crystallina Wood, 1815, p. 149; Hanley, 1846, p. 270, Pl. 57, fig. 43; Römer, 1872, p. 196, Pl. 38, figs. 1–2, Pl. 2, fig. 10.

DESCRIPTION: Shell small, thin, white, compressed, trigonal-ovate, subequilateral, inequivalve. Right valve is almost flat with only slight convexity, left valve is moderately convex; anterior end is slightly longer than posterior end and broadly rounded, posterior end is narrow with an acuminate extremity and bent to right. Dorsal margins are steep and slope down with equal angle on both sides, antero-dorsal margin is convex and postero-dorsal margin is concave; ventral margin is broadly convex and sinuous at the posterior end. Ligament is small and situated in a depressed ligamental groove, lunule is small and mostly on left valve; postero-umbonal fold is strong on right valve and weak on the left. Sculpture consists of widely spaced, thin, concentric ridges, the inter-spaces separating them are about five times as wide as the thickness of the ridges and are lined with very fine concentric striae. The hinge plate is thin and the cardinals are small; right anterior and left posterior cardinals are small and partly fused to the nymph, right posterior and left anterior cardinals are bifid and somewhat larger than the other cardinals. Right valve has two inequidistant and well developed laterals, the posterior one is situated below the distal end of ligamental groove, anterior one slightly closer to the cardinals than the posterior one; laterals of left valve are obsolete. Pallial sinus is large, its dorsal margin is posteriorly high and anteriorly descending until it touches the pallial line at a distance of one millimeter behind the anterior adductor muscle scar, ventral margin of it is entirely coalescent with pallial line. Anterior adductor muscle scar is larger than the posterior one and has elongate shape, posterior one is suborbicular.

FIGURED SPECIMEN: USNM 381903 (Barcelona Beach, Venezuela); length, 22.0 mm (100); height, 16.3 mm (74.2); thickness, 5.3 mm (29.2).

HABITAT: Coast of South Carolina to Venezuela.

Subgenus *Pinguitellina* Iredale

Pinguitellina Iredale, 1927, p. 76.
TYPE: *Tellina robusta* Hanley (by original designation).
GEOLOGIC RANGE: Recent.

Arcopagia (Pinguitellina) robusta (Hanley)
(Pl. 23, figs. 6–10)

Tellina robusta Hanley, 1844, p. 63; Hanley, 1846, p. 252, Pl. 56, fig. 23; Römer, 1871, p. 89, Pl. 24, fig. 406.

DESCRIPTION: Shell small, thin, trigonal-orbicular, white or yellowish in color, inflated, subequilateral, subequivalve, left valve more convex than right valve. Anterior end is slightly longer than posterior end and rounded, posterior is smoothly rounded; dorsal margins have equal and moderately steep slope on both sides; ventral margin is convex. Ligament is lanceolate, small, and situated in a depressed ligamental groove; lunule is short and mostly on the right valve. Sculpture consists of somewhat widely spaced, fine concentric, low ridges. Right anterior and left posterior cardinals are small, right posterior and left anterior cardinals are bifid and larger than the other cardinals. Right valve has two subequidistant, and well developed laterals, each one separated from the shell margin by a deep furrow, the posterior one is situated below the distal end of ligamental groove, anterior one is slightly closer to the cardinals than the posterior one; laterals of left valve are obsolete. Pallial sinus is large, its dorsal margin is descending, its anterior end is rounded and extends forward to within one millimeter of anterior adductor muscle scar, posterior half of its ventral margin is coalescent with pallial line. Adductor muscle scars are subequal and suborbicular.

FIGURED SPECIMEN: USNM 75589 (Tuamotu Islands); length, 9.7 mm (100); height, 7.8 mm (80.4); thickness, 3.8 mm (39.2).

HABITAT: Recorded from Philippines, Australia and South Pacific.

Subgenus *Punipagia* Iredale

Punipagia Iredale, 1930, pp. 397, 407.
TYPE: *Tellina subelliptica* Sowerby (= *Tellina hypelliptica* Salisbury) (by original designation).
GEOLOGIC RANGE: Recent.

Arcopagia (Punipagia) hypelliptica (Salisbury)

Tellina subelliptica 'Sowerby' Reeve, 1867, Pl. 39, figs. 220a, b (not of Meek and Hayden, 1856); Iredale, 1930, p. 397.

Tellina hypelliptica Salisbury, 1934, p. 89.

DESCRIPTION: I could not obtain a specimen of this species; therefore, I am quoting the original description: "Shell shortly suboval, thin, convex, white or purple, smooth, slightly striated near the margins; posterior side rather short, with radiating flexure rib-like on the right valve, dorsal margin convex, sloped, ventral margin a little excavated near the flexure, end subrotund; anterior side obliquely produced, ventral margin convex, umbones small, acute, ligament partly imbedded.—Sowerby."

HABITAT: Port Jackson, New South Wales.

REMARKS: Iredale makes the following statement about *T. hypelliptica:* "The hinge is similar to that of *Pinguitellina* as figured by me, but the pallial sinus differs as well as the texture."

GENUS LINEARIA CONRAD, 1860

DIAGNOSIS: Shell small to moderately large size, elongate-ovate, anterior and posterior ends rounded, dorsal margins moderately steep, ventral margin broadly arcuate. Beaks are small, umbones are subdued, postero-umbonal fold is absent or obsolete; ligamental groove is deep, narrow, and in general rather long. Sculpture, in most cases, is concentric, and in some it is both concentric and radial. There are two cardinals and two laterals in each valve. Pallial sinus is small to medium size, and its ventral margin is entirely or mostly free.

TYPE: *Linearia metastriata* Conrad (by monotypy).

Key to the Subgenera of Genus *Linearia*

A. Sculpture concentric and radial.
   a. Radial sculpture of strong ridges. . . . . *Linearia* s.s.
   aa. Radial sculpture of weak striae.
      b. Fine radial striae over entire shell surface. . . . . *Palaeomera*
      bb. Posterior area only with granulated radial striae. . . . . . . *Oene*
B. Sculpture concentric only.
   a. Beaks central or subcentral.
      b. Beaks centrally situated. . . . . *Tellinimera*
      bb. Beaks subcentral, slightly posterior to midlength. . . . . . . *Iredalesta*
                                        new subgenus
   aa. Beaks not central nor subcentral.
      c. Beaks situated posterior to mid-length.
         d. Pallial sinus small. . . . . *Arcopagella*
         dd. Pallial sinus of medium size. . . . . *Liothyris*
      cc. Beaks situated anterior to mid-length.
         e. Pallial sinus small. . . . . *Hercodon*
         ee. Pallial sinus of medium size.
            (1) Surface smooth. . . . . *Aenona*
            (2) Surface with irregular concentric sculpture. . . . . . . *Nelltia*

Note: The genus *Nelltia* Stephenson should probably be considered a subgenus of *Linearia.*

Subgenus *Linearia* Conrad

Linearia Conrad, 1860, p. 279.

TYPE: *Linearia metastriata* Conrad (by monotypy).

GEOLOGIC RANGE: Upper Cretaceous.

Linearia (Linearia) metastriata Conrad
(Pl. 24, figs. 12–15)

Linearia metastriata Conrad, 1860, p. 279, Pl. 46, fig. 7; Conrad, 1870, p. 73, Pl. 3, fig. 11; Whitfield, 1885, p. 165, Pl. 23, figs. 6–7; Stephenson, 1923, p. 329, Pl. 84, figs. 1–5.

DESCRIPTION: Shell small, thin, elongate-ovate, subequilateral, compressed. Anterior end is broadly rounded, posterior end is rounded and subtruncate. Beaks are low and situated slightly posterior of mid-length; dorsal margins are straight, and slope gently with equal angle on either side; ventral margin is broadly arcuate. Ligamental groove is narrow and deep, extending halfway along postero-dorsal margin. Surface has a decussate sculpture produced by concentric and radial ridges; former is strong on the middle of the shell and weak on both ends; latter is weak on the middle area but strong on both ends, and on the posterior end radial ridges are stronger than on the anterior end. There are two cardinals and two laterals in each valve; in the right valve both cardinals are long, slender, and obliquely directed forward and separated from each other by a deep and narrow socket; laterals are well developed, parallel to margin, inequidistant, the posterior one is more distant from cardinals than anterior one. In the left valve anterior cardinal is slender, long and obliquely directed forward; posterior cardinal is small, short, and directed vertically downward; laterals of left valve are weak. Pallial sinus is of medium size, slightly ascending, and ventral margin of it is entirely free. Adductor muscle scars are subequal.

FIGURED SPECIMEN: The specimens of this species deposited in the Museum are not in such a well-preserved condition as to be satisfactory for photography; therefore, the figures shown here are copied from Dr. Stephenson's illustrations. Measurements of a specimen of average size are: length, 19 mm; height, 12 mm.

OCCURRENCE: Eufaula, Alabama; Snow Hill calcareous member of Black Creek formation, N. C.; Woodbury clay (*Exogyra ponderosa* zone), near Haddonfield, N. J.; reported by Dr. Gardner from the Monmouth formation (*Exogyra costata* zone) in Prince George's County, Maryland; Ripley formation (*Exogyra costata* zone), Owl Creek, Ripley, Tippah County, Mississippi.

Subgenus *Tellinimera* Conrad

Tellinimera Conrad, 1860, p. 278.
Tellimera Conrad, 1870, p. 73.
TYPE: *Tellina eborea* Conrad (by subsequent designation Conrad, 1870).
GEOLOGIC RANGE: Cretaceous.

Linearia (Tellinimera) eborea (Conrad)
(Pl. 24, figs. 9–10)

Tellinimera eborea Conrad, 1860, p. 278, Pl. 46, fig. 14; Tryon, 1884, p. 169, Pl. 112, fig. 100; Whitfield, 1885, p. 164, Pl. 23, figs. 12–13.
Tellimera eborea Conrad, 1870, p. 73, Pl. 3, fig. 11.

DESCRIPTION: Shell small, thin, subtrigonal-ovate, compressed, equilateral. Anterior and posterior ends are rounded, the latter is slightly pointed; postero-umbonal fold is weak. Beaks are moderately prominent and centrally situated; dorsal margins are sloping rather steeply with equal angle on either side; the ventral margin is broadly arcuate. Sculpture consists of fine concentric striae. There are two cardinals, the anterior one is trigonal and directed slightly anteriorly, posterior one is bifid and oblique; posterior lateral is thick and longer than the anterior lateral.

REMARKS: In the collections of the Museum the only specimen of this species available is one where both valves are cemented together and it is impossible to open it without breaking; therefore, the description given here about the hinge character of this species is that given by Conrad. I was unable to observe the pallial sinus and I could not find any description of it.

FIGURED SPECIMEN: The illustrations of this species appearing in this work are a copy of those figured by Dr. Gardner (1916, pl. 42), and measurements are: length, 22.4 mm; height, 13.1 mm.

OCCURRENCE: Barbour County, Alabama.

Subgenus *Aenona* Conrad

Aenona Conrad, 1870, p. 74.

TYPE: *Tellina eufalensis* Conrad (=*Tellina eufaulensis* Conrad) (by subsequent designation Stoliczka, 1871).

GEOLOGIC RANGE: Upper Cretaceous.

Linearia (Aenona) eufaulensis (Conrad)
(Pl. 24, figs. 19–20)

Tellina eufaulensis Conrad, 1860, p. 277, Pl. 46, fig. 15.

Aenona eufalensis Meek, 1864, p. 14; Conrad, 1870, p. 74; Gardner, 1916, p. 697, Pl. 42, figs. 3, 4.

DESCRIPTION: Shell small, thin, somewhat compressed, elongate-subtrigonal, equivalve, inequilateral. Anterior end is short and broadly rounded, posterior end is acutely rounded; postero-umbonal fold is absent. Umbones are not prominent and they are situated slightly posterior of the mid-length; dorsal margins slope moderately with an equal angle on either side; ventral margin is broadly arcuate. Ligamental groove is short and narrow; lunule is very narrow, deep, and lanceolate. Surface is smooth except near ventral margin where it is marked by a few concentric striae. Hinge plate is narrow; in the right valve the anterior cardinal is short, thin, and lamellar; posterior cardinal is stout, short, bifid, and separated from the other cardinal by a wide, triangular socket; there are two well developed laterals situated equidistant from the cardinals. In the left valve, the anterior cardinal is trigonal and bifid; posterior cardinal is small and lamellar. Pallial sinus is of medium size, its anterior end extending forward to the mid-length of shell, and its ventral margin is free.

FIGURED SPECIMEN: The illustrations are a copy of those figured by Dr. Gardner (1916, pl. 42).

OCCURRENCE: Eufaula, Alabama; Monmouth formation, Oxon Hill, Prince George's County, Maryland; Woodbury clay, New Jersey; Ripley formation, Quitman County, Georgia; and Union and Tippah counties, Mississippi.

REMARKS: This type species was originally named *Tellina eufaulensis* by Conrad in 1860; later, in 1870, under the generic name of *Aenona*, he listed two species, the first one of which is *A. eufalensis*, and reference is given to his 1860 work with plate and figure numbers in which he has named the species after the geographic name Eufaula. This shows without doubt that *Eufalensis* is a misprint of the original name *eufaulensis*. Although Stoliczka has adopted the incorrect name subsequent to the original one, I believe it will be proper to drop the misprinted name of 1870 and retain the original name *eufaulensis*.

Subgenus *Palaeomoera* Stoliczka

Palaeomoera Stoliczka, 1870, p. 116.

TYPE: *Tellina strigata* Goldfuss (by subsequent designation, Stoliczka, 1871).
GEOLOGIC RANGE: Cretaceous.

Linearia (*Palaeomoera*) *strigata* (Goldfuss)
(Pl. 24, fig. 11)

Tellina strigata Goldfuss, 1826, p. 224, Pl. 147, fig. 18.

DESCRIPTION: The description given by Stoliczka is as follows: "Shell elongated, hinder part shorter, the upper declivity slightly convex, posterior end sub-truncate, beaks directed forward, ligament situated on thickened but not prominent fulcra; hinge with one anterior, long lamelliform tooth in each valve, bifid in right, single in the left valve; posterior cardinal teeth not distinctly traceable in either valve; laterals less distinct in the left valve."

I was unable to obtain a specimen of this species; however, from the observations on the illustrations I may add the following additional description: Shell of medium size, elongate-ovate, subequilateral, umbones moderately prominent and situated slightly posterior of mid-length. Anterior end is rounded, posterior end is bluntly truncate; postero-dorsal margin is slightly convex, antero-dorsal margin is straight, ventral margin is broadly arcuate; postero-umbonal fold is moderate. Sulupture consists of both concentric and radial fine striae together with some widely spaced growth lines.

FIGURED SPECIMEN: The illustration appearing in this work is a copy of original figure, and the measurements are: length, 35.5 mm; height, 21.6 mm.

OCCURRENCE: Cretaceous of Germany.

Subgenus *Arcopagella* Meek

Arcopagella Meek, 1871, p. 308.
TYPE: *Arcopagella mactroides* Meek (by monotypy).
GEOLOGIC RANGE: Dakota Sandstone, Cretaceous.

Linearia (*Arcopagella*) *mactroides* (Meek)
(Pl. 24, figs. 3–4)

Arcopagella mactroides Meek, 1871, p. 309, figs. A, B.

DESCRIPTION: Shell small, subelliptical, moderately convex, subequilateral, equivalve. Anterior and posterior ends are narrowly rounded, posterior end is narrower than anterior particularly below where there seems to be a slight tendency to a flexure; ventral margin is broadly convex. Dorsal margins are sloping moderately and almost equally on either side; antero-dorsal margin is convex, and postero-dorsal margin slightly concave. Beaks are situated slightly anterior of mid-length, ligamental groove is narrow and external, and surface of the shell is smooth. There are two cardinals and two laterals in each valve. On the right valve cardinals are slender and divergent, the anterior one is more oblique than the other and nearly connected with anterior lateral. Both of the laterals are elongate and parallel to the margin. On the left valve the anterior cardinal is direct, large, trigonal and bifid; posterior cardinal is slender and oblique; laterals of left valve are similar to those of the right valve. Pallial sinus is small, slightly ascending; its anterior end is broadly rounded and extending forward just short of mid-length of the shell, and about half of its ventral margin is free. Adductor muscle scars are subequal, anterior one is elongate and posterior one suborbicular.

FIGURED SPECIMEN: In the collections of the museum there is only one valve of this species; therefore, the figures appearing in this work are a copy of those originally illustrated by Meek. Measurements are as follows: length, 19.8 mm (100); height, 13.5 mm (68.2); thickness, 6.6 mm (33.4).

OCCURRENCE: Twelve miles southwest of Salina, Salina County, Kansas.

Subgenus *Hercodon* Conrad

Hercodon Conrad, [1873], p. 10.
TYPE-SPECIES: *Hercodon ellipticus* Conrad (by monotypy).
GEOLOGIC RANGE: Upper Cretaceous.

Linearia (Hercodon) ellipticus (Conrad)
(Pl. 23, figs. 11–14)

Hercodon ellipticus Conrad, in Kerr, 1875, p. 10, Pl. 2, figs. 2 and 8 (not the small outline figure).
Tellina elliptica Stephenson, 1923, p. 322, Pl. 81, figs. 1–4, Pl. 82, figs. 7–8.

DESCRIPTION: Shell of moderately large size, rather thick, subelliptical in outline, depressed convex, equivalve, inequilateral. Anterior and posterior ends are somewhat narrowly rounded, the latter is slightly pointed; dorsal margins have an equal and moderate slope, and ventral margin is broadly arcuate. Beaks are small, and situated about 0.44 of the length of the shell from anterior end. Ligamental groove is deep, narrow and extending about a third of the distance to posterior end; postero-umbonal fold is absent. Sculpture consists of irregular, fine, concentric ridges, and very fine radial striae. There are two cardinals and two laterals in each valve. On the right valve the posterior cardinal is large, trigonal and slightly oblique; the anterior cardinal is short, oblique, and smaller than the other cardinal. The anterior lateral of right valve is short, prominent, partly fused with anterior cardinal, and not quite parallel to the margin; right posterior lateral is short, weakly developed, and situated just below the posterior extremity of ligamental groove. Anterior cardinal of left valve is large, nearly direct, and bifid; the posterior cardinal is thin, lamellar, and oblique. Left posterior lateral is obsolete, and situated at the distal end of ligamental groove; left anterior lateral is short, low, poorly developed, and situated immediately in front of the anterior cardinal socket. Pallial sinus is characteristically short, its dorsal margin is straight, its anterior end extending less than the mid-length of the shell, and about half of its ventral margin is free. Within the shell, a slight rib extends just below the beak to the back of anterior adductor scar. Anterior adductor muscle scar is elongate and slightly smaller than the posterior one which has suborbicular shape.

FIGURED SPECIMEN: The specimens of this species in the collections of the National Museum are not in such a well preserved state as to be usable for photographing; therefore, I copy Dr. Stephenson's illustrations which show the details quite clearly. The dimensions are given as: length, 47 mm (100); height, 27 mm (57.5); thickness, 6 mm (12.9).

OCCURRENCE: Snow Hill calcareous member of Black Creek formation (upper part of *Exogyra ponderosa* zone), Snow Hill, Greene County, North Carolina.

Subgenus *Liothyris* Conrad

Liothyris Conrad, [1873], p. 10 (not Douville, 1879).
TYPE: *Linearia carolinensis* Conrad (by monotypy).
GEOLOGIC RANGE: Upper Cretaceous.

Linearia (Liothyris) carolinensis Conrad
(Pl. 24, figs. 5–8)

Linearia carolinensis Conrad, in Kerr, 1875, p. 9, Pl. 1, figs. 20, 23, 24; Stephenson, 1923, p. 326, Pl. 83, figs. 4–12.

DESCRIPTION: Shell of moderately large size, rather thick, elongate-ovate, slightly convex. Posterior end is bluntly truncate, anterior end is acutely rounded and slightly narrower than posterior end. Beaks are moderately prominent, orthogyrate, and situated about two-fifths of the length of the shell from anterior end. Antero-dorsal margin is straight and sloping with a moderate angle; postero-dorsal margin is slightly convex and sloping rather steeply; ventral margin is broadly convex. There is an obsolete postero-umbonal fold; ligamental groove is deep, narrow and extending along half the length of postero-dorsal margin; lunule and escutcheon are narrow and sharply defined. The surface is somewhat smooth and marked by fine concentric striae some of which are rather strong. In the large individuals the hinge plate is somewhat heavy; the right anterior cardinal is small, thin, and directed obliquely forward and downward; right posterior cardinal is short, thick, directed vertically downward, and separated from anterior cardinal by a large and deep socket; on the left valve the anterior cardinal is prominent and directed obliquely forward and downward, the posterior cardinal is small and obsolete. In the young and small sized individuals the hinge plate is narrow; right valve has two slender, long, and very oblique cardinals directed forward and downward, and separated by a narrow socket; left valve has a slender and rather long anterior cardinal and a small obsolete posterior cardinal. Laterals in both large and small individuals are similar, consisting of two well-developed inequidistant teeth on the right valve, the posterior one is situated below the distal end of ligamental groove, and anterior one is slightly closer and less distant from cardinals. Laterals of left valve are obsolete. Internally a radial rib extends from below the beaks to the posterior margin of anterior adductor muscle scar. Pallial sinus is of medium size, ascending, its anterior end acutely rounded and extending to the middle of the shell, and its ventral margin entirely free. Anterior adductor muscle scar is elongate and slightly larger than the posterior one that is of orbicular shape.

FIGURED SPECIMEN: The specimens of *carolinensis* present in the collections of the Museum were not well enough preserved to show all the details in photography; therefore, the figures are copied from Dr. Stephenson's illustrations. Measurements are: length, 45.0 mm (100); height, 30.0 mm (66.7); thickness, 6.5 mm (14.5).

OCCURRENCE: Snow Hill calcareous member of Black Creek formation (upper part of *Exogyra ponderosa* zone); Snow Hill, Greene County, North Carolina.

Subgenus Oene Conrad

Oene Conrad, [1873], p. 9.
TYPE: *Oene plana* Conrad (by monotypy).
GEOLOGIC RANGE: Upper Cretaceous.

Linearia (Oene) plana Conrad
(Pl. 24, figs. 16–18)

Oene plana Conrad, in Kerr, 1875, p. 9.
Linearia plana Stephenson, 1923, p. 327, Pl. 83, figs. 1–3.
DESCRIPTION: Shell small, thin, elongate-subtrigonal, compressed, inequivalve, equilateral. Anterior end is rounded; posterior end is bluntly truncate, narrower than anterior end, and slightly bent to right. Beaks are small and centrally situated; dorsal margins are straight and slope steeply with equal angle on either side; ventral margin is mostly straight but slightly concave in the middle. Postero-umbonal fold is of moderate strength on the left valve, and weak on the right valve; ligamental groove is deep, narrow and short, extending only a third of the distance to posterior end; lunule is deep, rather long and very narrow; escutcheon is not defined. Sculpture, in front of

postero-umbonal fold, consists of fine concentric ridges; dorsal area on the left valve is marked by fine radial lines crossed by sharp growth lines, on the right valve the posterior area is marked by very fine radial lines. Hinge plate is very narrow; in the left valve the anterior cardinal is long, well developed, obliquely directed forward; posterior cardinal is small, short, and directed vertically downward. Left anterior lateral is obsolete and situated only a short distance from anterior cardinal, posterior lateral is elongate, obsolete, and distant from cardinals. I have not seen the right valve of this species, but it is described as having two well developed somewhat long cardinals directed obliquely forward, the anterior one slender and the posterior one thick. Laterals of the right valve have not been mentioned in previous descriptions, but judging from the fact that all of the other members of Tellinidae have strong laterals on the right, while those of the left valve are relatively weak, I will venture to consider it probable that in this species also the laterals of right valve corresponding to those of left valve are relatively strong.

FIGURED SPECIMEN: The figures are a copy of those illustrated by Dr. Stephenson; dimensions of the left valve are: length, 22 mm (100); height, 11 mm (50.0); thickness, 2 mm or 4 mm for the shell (8.4).

OCCURRENCE: Snow Hill calcareous member of Black Creek formation (upper part of *Exogyra ponderosa* zone), Snow Hill, Greene County, North Carolina.

Subgenus *Iredalesta*, New Subgenus

DIAGNOSIS: Shell small, subelliptical in outline, moderately convex, subequilateral. Beaks are somewhat prominent, small, and situated slightly posterior to the mid-length. Ligamental groove is short and deep; postero-umbonal fold is absent, and sculpture is concentric. There are two cardinals and two laterals in each valve. Within the shell there are three feeble ribs; pallial sinus is short, horizontal, and about half of its ventral margin is free.

TYPE: *Tellina stephensoni* Salisbury.

GEOLOGIC RANGE: Upper Cretaceous.

Linearia (*Iredalesta*) *stephensoni* (Salisbury)
(Pl. 24, figs. 1–2)

Tellina simplex Stephenson, 1923, p. 324, Pl. 82, figs. 4–6.
Tellina stephensoni Salisbury, 1934, p. 89.

DESCRIPTION: Shell small, subelliptical in outline, moderately convex, subequilateral. Anterior and posterior ends are sharply rounded, posterior margin more acutely than anterior; dorsal margins have equal and moderately steep slopes; ventral margin is broadly arcuate. Beaks are somewhat prominent and situated slightly posterior to the mid-length; ligamental groove is short and deep; postero-umbonal fold is absent. Sculpture consists of fine concentric ridges, some of them become rather coarse near margins; there are also very fine radial striae that are observable only by the aid of lens. On the right valve the posterior cardinal is large, bifid, and slightly oblique; the anterior cardinal is short, small, oblique and partly fused with the anterior lateral. Right anterior lateral is short, prominent, partly fused with anterior cardinal, and the hinge plate in front of it is very short; posterior lateral of right valve is short, weakly developed and situated just below the posterior end of ligamental groove. Left anterior cardinal is direct, large, and bifid; posterior of left valve is weakly developed, slender and oblique. On the left valve the anterior lateral is approximate, short and moderately large; the posterior lateral is obsolete. Pallial sinus is short, its dorsal margin is horizontal, anterior end rounded and extending forward just short of mid-length of the shell, and half of its ventral margin is free. Within the shell there are three feeble radial ribs extending from below the beak, one in front of the posterior adductor muscle scar, another just in front of pallial sinus, and the third just back of anterior adductor scar. Anterior adductor muscle scar is elongate and slightly smaller than the posterior one which has suborbicular shape.

REMARKS: This species is distinguished from *H. ellipticus* in that it is smaller in size, has three internal ribs, and the anterior end is slightly longer than the posterior; whereas, *H. ellipticus* is large, has only one internal rib, and its anterior end is shorter than the posterior end.

FIGURED SPECIMEN: The specimen to which I had access was not in a well-preserved state and for that reason the figures which appear in this work are copied from Dr. Stephenson's excellent illustrations. Measurements are as follows: length, 25 mm (100); height, 14 mm (56.0); thickness, 3 mm (12.0).

OCCURRENCE: Snow Hill calcareous member of Black Creek formation (upper part of *Exogyra ponderosa* zone), Snow Hill, Greene County, North Carolina.

GENUS STRIGILLA TURTON, 1822

DIAGNOSIS: Shell small to very small size, trigonal-orbicular, white with rose or brown coloring, inequilateral, inflated. Both ends are rounded, the umbones are situated anterior to the mid-length, postero-umbonal fold is obsolete. Sculpture consists of finely incised rugae which are oblique in the middle of the shell but divaricate on the posterior end, and concentric or divaricate on the anterior end. There are two cardinals in each valve; two strong laterals on the right valve and two obsolete laterals on the left valve. Pallial sinus may be large, small, or discrepant in different species, but always the ventral margin is coalescent with the pallial line.

TYPE: *Tellina carnaria* Linnaeus (by subsequent designation Stoliczka, 1871).

GEOLOGIC RANGE: Eocene to Recent.

Key to the Subgenera of Genus *Strigilla*

A. Pallial sinus not discrepant in two valves.
    a. Anterior end of pallial sinus not coalescent with anterior adductor
        muscle scar and surface of the shell without smooth area.
        b. Incised striae on the posterior area with only one inflection,
            and terminating in an upward trend—generally larger than 10
            mm. . . . . *Strigilla* s.s.
        bb. Incised striae on the posterior area with two inflections and
            terminating in a downward trend—smaller than 10 mm. . . . *Pisostrigilla*
        bbb. Incised striae strongly oblique crossing the shell from one mar-
            gin to the other but without inflection—smaller than 10 mm. *Simplistrigilla*
    aa. Anterior end of pallial sinus coalescent with the anterior adductor
        muscle scar and with smooth area on the surface.
        c. Smooth area on the anterior part of left valve only—generally
            larger than 10 mm. . . . . *Rombergia*
        cc. Smooth area on the anterior part of both valves—smaller than
            10 mm. . . . . *Roemerilla*
                                        new subgenus

B. Pallial sinus discrepant in two valves. . . . . *Aeretica*

NOTE: To the subgenera in this key must be added *Simplistrigilla* Olsson, to be found in the Appendix.

Subgenus *Strigilla* Turton

Strigilla Turton, 1822, p. 117.
Strigella Gray, 1842, p. 91.
Limicola 'Leach' Gray, 1852, p. 296 (not Koch, 1816).
Strigula Menke & Pfeiffer, 1861, Vol. VII Index.
Strigillina Stoliczka, 1870, p. 120 (not Dunker, 1862).
TYPE-SPECIES: *Tellina carnaria* Linnaeus (by subsequent designation Gray, 1847).
GEOLOGIC RANGE: Recent.

Strigilla (Strigilla) carnaria (Linnaeus)
(Pl. 25, figs. 1–5)

Tellina carnaria Linnaeus, 1758, p. 676; Römer, 1872, p. 183, Pl. 26, figs. 7–9.
Cardium carneosum Da Costa, 1779, p. 181.
Lucina carnaria Lamarck, 1818, p. 541.
Strigilla carnaria Turton, 1822, p. 117, Pl. 7, fig. 15.

DESCRIPTION: Shell small, solid, inflated, trigonal-orbicular, white with rose coloring, inequivalve, inequilateral. Right valve is slightly more convex than the left valve; anterior end is broadly rounded and shorter than the posterior end which is acutely rounded. Umbones are situated about a third of the length from anterior end; postero-umbonal fold is obsolete on the right valve but is not perceptible on the left valve. Dorsal margins slope steeply, ventral margin is convex. Ligament is small and situated in a depressed ligamental groove; a small lunule is present. Sculpture consists of fine incised rugae which are oblique on the central area of the shell but divaricant on the anterior and posterior ends. On the anterior end the incised rugae have a moderate flexure, but on the posterior area they have a sharp inflection and terminate with an upward trend at the posterior margin. Right posterior cardinals are large, trigonal, and bifid; right anterior cardinal is small and pointed. Left anterior cardinal is bifid and somewhat small, left posterior cardinal is lamellar, thin, and partly fused with the nymph. Right valve has two strong, inequidistant laterals which are separated from the margin by a furrow. Left valve has two obsolete laterals. Pallial sinus is small, its dorsal margin is descending, its anterior end is rounded and 6 millimeters from anterior adductor muscle scar, the ventral margin of it is entirely coalescent with the pallial line. Anterior adductor muscle is elongate, the posterior one is of suborbicular shape.

FIGURED SPECIMEN: USNM 27668 (St. Croix, Virgin Islands); length, 27.0 mm (100); height, 23.9 mm (88.5); thickness, 12.9 mm (47.8).

HABITAT: From Florida through Caribbean to north coast of Brazil.

Subgenus *Pisostrigilla* Olsson

Pisostrigilla Olsson, 1961, Panamic-Pacific Pelecypoda, p. 390.
TYPE-SPECIES: *Tellina pisiformis* Linnaeus (by original designation).
GEOLOGIC RANGE: Oligocene to Recent.

Strigilla (Pisostrigilla) pisiformis (Linnaeus)
(Pl. 25, figs. 6–10)

Tellina pisiformis Linnaeus, 1758, p. 677; Römer, 1872, p. 188, Pl. 37, figs. 1–3.
Cardium discors Montagu, 1803, p. 84.
Strigilla pisiformis Turton, 1822, p. 119.
Lucina pisiformis Thorpe, 1844, p. 75.

DESCRIPTION: Shell very small, solid, white with rose coloring on the umbones, trigonal-orbicular, inflated, inequilateral, equivalve. Anterior end is broadly rounded and shorter, posterior end rounded and slightly narrow; the ventral margin is convex. Umbones are situated anterior to the mid-length, dorsal margins are sloping steeply, and postero-umbonal fold is absent. A small ligament is situated in a depressed ligamental groove, and there is a small lunule. Sculpture consists of fine incised rugae which are oblique on the central area of the shell but divaricant on both ends, and there are some fine concentric growth lines. The incised rugae, which form a major part of the sculpture, bend with a moderate flexure on the anterior area, whereas on the posterior area they make two sharp inflections and terminate at the posterior margin with a downward trend. Right posterior cardinal is bifid, trigonal, and relatively large; right anterior cardinal is small. On the left valve the anterior cardinal is small and bifid, the posterior one is lamellar, thin and partly fused with the nymph. Right valve has two inequidistant strong laterals; laterals of left valve are obsolete. Pallial sinus is large, its dorsal margin is descending until it touches the pallial line one millimeter behind the anterior adductor muscle scar; ventral margin of it is entirely coalescent with pallial line. Anterior adductor muscle scar is elongate, the posterior one is orbicular.

FIGURED SPECIMEN: USNM 28073 (Samana Bay, Dominican Republic); length, 8.8 mm (100); height, 8.2 mm (93.2); thickness, 5.1 mm (58.0).

HABITAT: Florida Keys and West Indies.

REMARKS: *S. pisiformis* L. is distinguished from *S. carnaria* L. by the fact that the former is smaller in size, has larger pallial sinus relative to size, and the incised rugae which form the sculpture make two inflections on the posterior area and terminate with a downward trend at the posterior margin. In contrast, *S. carnaria* L., other than being larger in size, has a smaller pallial sinus relative to its size, has an obsolete postero-umbonal fold on the right valve, and the incised rugae make only one inflection on the posterior area and terminate with an upward trend at the posterior margin.

Subgenus *Rombergia* Dall

Rombergia Dall, 1900b, p. 1038.

TYPE-SPECIES: *Strigilla rombergi* Mörch (by original designation).

GEOLOGIC RANGE: Recent.

Strigilla (Rombergia) rombergi Mörch
(Pl. 25, figs. 11–15)

Strigilla rombergi Mörch, 1852, p. 157.

Tellina rombergi Deshayes, 1854, p. 355; Römer, 1872, p. 187, Pl. 36, figs. 13–16.

DESCRIPTION: Shell small, solid, white with suffused rose color, inflated trigonal-orbicular, inequilateral, subequivalve. Anterior end is broadly rounded and shorter, posterior end rounded and somewhat narrow, the ventral margin is convex. Umbones are situated one-third of the length from the anterior end, dorsal margins are sloping steeply and the postero-umbonal fold is absent. Ligament is small and situated in a depressed ligamental groove, and there is a small lunule. Sculpture is discrepant on two valves by the presence of a smooth area on the anterior part of left valve extending from umbone to ventral margin. The sculpture consists of fine incised rugae which are oblique on the central area of the shell, have a moderate flexure on the anterior area, but they have a sharp inflection on the posterior margin. Right posterior cardinal is trigonal and bifid, and right anterior cardinal is small. On the left valve the anterior cardinal is bifid and relatively small, the posterior cardinal is lamellar, thin and partly fused with the nymph. Right valve has two strong, inequidistant laterals which are separated from the margin by furrow; laterals of the left valve are obsolete. Pallial sinus is large, its dorsal margin descending, its anterior end is narrow and coalescent with the lower end of anterior adductor muscle scar, and its ventral margin is entirely coalescent with pallial line. Anterior adductor muscle scar is elongate, and posterior one is orbicular.

FIGURED SPECIMEN: USNM 162985 (Jeremie, Haiti); length, 21.4 mm (100); height, 19.8 mm (92.5); thickness, 9.6 mm (45.0).

HABITAT: West Indies to north coast of Brazil.

Subgenus *Roemerilla,* New Subgenus

DIAGNOSIS: Shell very small, trigonal-orbicular, inflated, white with rose coloring mostly on the umbonal area, inequilateral, subequivalve. Both ends are rounded, the anterior end is broad and shorter. Postero-umbonal fold is absent, ligament is small and there is a small lunule. Sculpture consists of fine, incised rugae which are oblique on the central area of shell, and divaricant on both ends; on the posterior area the rugae have only one inflection and terminate with a downward trend. There are two cardinals in each valve, two well developed laterals on the right valve, and two obsolete laterals on the left valve. Pallial sinus is large, its anterior end is coalescent with the anterior adductor muscle scar, and its ventral margin is entirely coalescent with pallial line.

TYPE: *Strigilla cicercula* Philippi.

GEOLOGIC RANGE: Recent.

Strigilla (Roemerilla) cicercula (Philippi)
(Pl. 25, figs. 16–20)

Tellina cicercula Philippi, 1846, p. 19.
Tellina dichotoma Philippi, 1846, p. 20.
Tellina ervilia Philippi, 1846, p. 20.
Strigilla interrupta Mörch, 1860, p. 189.
Strigilla maga Mörch, 1860, p. 189.
Tellina maga Römer, 1872, p. 189, Pl. 37, figs. 4–6.
Strigilla cicercula Dall, 1900, p. 305.

DESCRIPTION: Shell very small, rather thin, solid, trigonal-orbicular, white with rose color mostly on the umbonal area, inequilateral, subequivalve. Anterior end is broadly rounded and shorter. Posterior end rounded and somewhat narrow. Dorsal margins are steep, ventral margin is convex and posterior-umbonal fold is absent. Ligament is small and situated in a depressed ligamental groove; there is a small lunule. Sculpture consists of fine incised rugae which are oblique on the central area of the shell, they have a moderate flexure on the anterior area, but on the posterior area they have a sharp inflection and terminate with a downward trend at the posterior margin. There is a smooth area on each valve extending from umbone to the antero-ventral margin. Right posterior and left anterior cardinals are bifid, right anterior cardinal is pointed and small; left posterior cardinal is lamellar, thin, and partly fused with the nymph. Right valve has two strong inequidistant laterals which are separated from the margin by a furrow; laterals of left valve are obsolete. Pallial sinus is large, its dorsal margin is posteriorly high and anteriorly descending, its anterior end coalescent with the anterior adductor muscle scar, and its ventral margin is entirely coalescent with pallial line. Anterior adductor muscle scar is elongate, the posterior one is suborbicular.

FIGURED SPECIMEN: USNM 73695 (Acapulco, Mex.); length, 8.9 mm (100); height, 7.7 mm (86.5); thickness, 5.1 mm (57.4).

HABITAT: Gulf of California to Ecuador.

REMARKS: The above description is that of two specimens which are from Philippi's original collection; because they are two valves of different size the figures were made not from these specimens but from a specimen identical with the cotypes in every respect.

Subgenus *Aeretica* Dall

Aeretica Dall, 1900, p. 1038.
TYPE-SPECIES: *Strigilla senegalensis* Hanley (by original designation).
GEOLOGIC RANGE: Recent.

Strigilla (Aeretica) senegalensis (Hanley)
(Pl. 25, figs. 21–25)

Tellina senegalensis Hanley, 1844, p. 68; Hanley, 1846, p. 259, Pl. 56, fig. 17; Römer, 1872, p. 194, Pl. 37, figs. 17–19.

DESCRIPTION: Shell small, solid, white with concentric bands of brown color, inflated, trigonal-orbicular, inequilateral, inequivalve. Anterior end is broadly rounded, posterior end is rounded and somewhat narrow. Right valve is more convex than the left valve, its beak stands slightly higher and its antero-dorsal margin overlaps that of the left valve. Umbones are situated about one third of the length from the anterior end. Dorsal margins are sloping steeply, ventral margin is convex and slightly sinuous at the posterior part. Postero-umbonal fold is obsolete on the right valve and absent on the left valve. Ligament is small and situated in a depressed ligamental groove. Sculpture consists of fine incised rugae which are concentric on the anterior third of the shell, oblique on the remaining part of the shell having a sharp inflection on the posterior area and terminating with an upward trend at the posterior margin. Right posterior and left anterior cardinals are bifid and relatively large, right anterior cardinal is small and pointed; left posterior cardinal is small, lamellar, thin, and partly fused with the nymph. Right valve has two strong, inequidistant laterals which are separated from the margin by a furrow; laterals of left valve are obsolete. Pallial sinus is discrepant in two valves; that of right valve is smaller, anterior part of its dorsal margin is descending steeply until it touches the pallial line five millimeters behind the anterior adductor muscle scar, and its ventral margin is entirely coalescent with pallial line. Pallial line of left valve is large, its dorsal margin is only slightly descending, its anterior end is coalescent with the lower end of anterior adductor muscle scar, and its ventral margin is entirely coalescent with pallial line. Anterior and posterior adductor muscle scars are suborbicular, and the latter is slightly smaller.

FIGURED SPECIMEN: USNM 6805 (Coast of Senegal); length, 22.3 mm (100); height, 19.7 mm (88.4); thickness, 9.6 mm (43.0).

HABITAT: Coast of Senegal, Africa.

GENUS FINLAYELLA LAWS, 1933

Finlayella Laws, 1933, p. 319.
DIAGNOSIS: Shell small, fragile, anterior end is rounded, posterior end strongly angled, and bent to right. Beaks are slightly posterior to mid-length, antero-dorsal margin is sloping moderately, postero-dorsal margin is steep. Right valve has two cardinals and thin lateral teeth, the anterior lateral closer to cardinals; left valve has a narrow cardinal. Pallial sinus is discrepant in the two valves, that of right valve being larger; the ventral margin of pallial sinus is entirely free.

TYPE: *Finlayella sinuaris* Laws (by original designation).
GEOLOGIC RANGE· Upper Oligocene.

Finlayella sinuaris Laws
(Pl. 26, figs. 1–3)

Finlayella sinuaris Laws, 1933, p. 319, Pl. 29 figs. 2, 3, 6.

DESCRIPTION: Following is the original description of this species, of which I was unable to obtain a specimen: "Shell small, fragile, well rounded over anterior end, strongly angled posteriorly. Beaks a little posterior. Antero-dorsal margin slightly curved; postero-dorsal one straight, descending more rapidly than antero-dorsal one. Posterior end bent to right. Sculpture of fine sharp, dense concentric growth lines, and in addition four or five sharp, distant lamella-like threads, regularly spaced and developed all around margins, and converging to beaks. Posterior flattened area carries several low, wide radial undulations. Right hinge wider in front than behind; it has two triangular cardinals, the posterior one bifid; the cardinals separated by a deep pit immediately below umbo; lateral teeth thin and sharp, anterior one arises nearer umbo than does posterior one. Left hinge with a narrow, vertical cardinal equally bifid and with wide triangular depressions on each side; hinge-plate not differentiated from margins of valves. Pallial sinus of right valve wide, sharply rounded at extremity, reaching a little over halfway across valve; that of left valve wide, very deep, bluntly rounded at extremity, reaching almost to anterior adductor scar. Valve margins smooth."

"Height, 7.0 mm; length, 9.2 mm; thickness (one valve), about 1.2 mm."

"LOCALITIES: Sutherlands, Tengawai River, South Canterbury (type and several paratypes); White Rock River, South Canterbury, New Zealand (a single left valve). These are Awamoan horizons."

REMARKS: *Finlayella sinuaris* Laws appears to be very closely related to *Arcopagia;* because in the description only one cardinal is mentioned on the left valve, I leave it as a separate genus, but if later examination of *sinuaris* Law reveals an obsolete left posterior lateral, *Finlayella* should be placed under *Arcopagia* with subgeneric rank.

GENUS TELLIDORA H. & A. ADAMS, 1856

DIAGNOSIS: Shell of medium size, white, rather thin, trigonal, compressed, inequivalve, subequilateral. Beaks are situated slightly anterior to mid-length, dorsal margins slope down with almost equal angle, and ventral margin is arcuate. Both lunule and escutcheon are prominent, and a small ligament is situated in a depressed ligamental groove. Sculpture is concentric, and dorsal margins are spinose. There are two cardinals in each valve, two strong laterals in the right valve, and laterals of left valve are obsolete. Pallial sinus is large and more than half of its ventral margin is free.

Key to The Subgenera of Genus *Tellidora*

A. Inequivalve. . . . . *Tellidora* s.s.
B. Equivalve. . . . . *Tellipiura*

Subgenus *Tellidora* H. & A. Adams, 1856 sensu stricto

Tellidora "Mörch" H. & A. Adams, 1856, p. 401.

TYPE: *Tellina burneti* Broderip and Sowerby (by subsequent designation Stoliczka, 1870).

GEOLOGIC RANGE: Recent.

Tellidora (Tellidora) burneti (Broderip and Sowerby)
(Pl. 26, figs. 4–8)

Tellina burneti Broderip and Sowerby, 1829, p. 362, Pl. 9, fig. 2; Hanley, 1846, p.

271, Pl. 58, fig. 99; Reeve, 1867, Pl. 35, figs. 199a, b; Römer, 1872, p. 198, Pl. 38, figs. 6–9.

DESCRIPTION: Shell of medium size, white, rather thin, trigonal, compressed, inequivalve, subequilateral. Anterior end is slightly shorter than the posterior, lunule and escutcheon are prominent such that each one extends along the entire length of dorsal margin, and a short ligament is situated in a depressed ligamental groove. Dorsal margins are sloping almost equally with a steep angle, and they are strongly spinous. Antero-dorsal margin is strikingly concave, whereas postero-dorsal margin has only a slight concavity; ventral margin is convex and moderately sinuous at the posterior end. Right valve is flat except at the posterior end where it is slightly bent to right, left valve is moderately convex. Beaks are prosogyrate and the beak of left valve bends over that of right valve; postero-umbonal fold is obsolete. Sculpture consists of fine concentric ridges with interspaces in general having a width three times that of the ridges; however, the interspaces do not possess uniformity in the width, occasionally some are wider than the rest. In some of the smaller individuals of this species the sculpture is irregular and obsolete close to posterior end. There are two cardinals in each valve; right posterior and left anterior cardinals are rather long, slender, and bifid; right anterior and left posterior cardinals are small. Laterals of right valve are strong, distant and situated at the mid-length of dorsal margins; laterals of left valve are obsolete. Pallial sinus is large, and in some individuals it is slightly discrepant, that of left valve being somewhat larger. Dorsal margin of pallial sinus is ascending on the posterior part and descending on the anterior part, its anterior end is acutely rounded and four millimeters from anterior adductor muscle scar; slightly more than half of its ventral margin is free. Anterior adductor muscle scar is elongate, posterior one is suborbicular.

FIGURED SPECIMEN: USNM 2551 (Mazatlan, Mexico); length, 39.1 mm (100); height, 33.3 mm (85.0); thickness, 6.4 mm (16.3).

HABITAT: Lower California to Ecuador.

Subgenus *Tellipiura* Olsson

Tellipiura Olsson, 1944, p. 63.

TYPE-SPECIES: *Tellidora* (*Tellipiura*) *peruana* Olsson (by original designation).

GEOLOGIC RANGE: Cretaceous (Upper Senonian).

Tellidora (*Tellipiura*) *peruana* Olsson
(Pl. 26, figs. 9–12)

Tellidora (*Tellipiura*) *peruana* Olsson, 1944, p. 63, Pl. 6, figs. 13–16.

COMMENT: Through the kindness of Dr. Olsson I received some fossil specimens of *T. peruana;* however, I found that his description gives more details than it is possible to observe from the specimens, therefore I am giving his original description:

DESCRIPTION: "Shell of medium size, equivalve, subtrigonal to subrounded in outline, the length being slightly greater than the height; the valves appear to be nearly flat but in reality are slightly convex as may be seen when viewed from above; the hinge margins, particularly on the posterior side, are strongly dentated by a series of sharp, tooth-like projections similar to those found on Recent species of *Tellidora;* the beaks are small, sharply pointed and curved forward, there being no clearly differentiated umbones; the sculpture is formed by a series of regularly spaced, concentric riblets which are persistent over the whole shell and overrun by a weakly differentiated posterior-dorsal area either by a smoothing of the sculpture or by a faint furrow; in the left valve this area is slightly flexed.

"Length, 21.5 mm; height, 22.5 mm; diameter, 4.5 mm."

"Length, 26 mm; height, 19.5 mm; diameter ——————."

"A rather poor impression of the hinge of the left valve shows the dentition and the immersed ligament to be essentially similar to that of *Tellidora burneti* Broderip and Sowerby of the West Coast Recent. The latter it also resembles closely by its shape, toothed dorsal margins and the slight but distinct posterior-dorsal flexur."

"The *Astarte*-like form and strong sculpture is also similar to that of *Astartemya* Stephenson (Stephenson, E. W.: The larger invertebrate fossils of the Navarro group of Texas, The University of Texas Pub. No. 4101, p. 173; pl. 27, figs. 1–3) recently described from the Navarro group of Texas but differs in shape and by its toothed margins."

FIGURED SPECIMENS: I have used those illustrated by Dr. Olsson in his original description of this species.

OCCURRENCE: Lower fossil zone, Tortuga, Peru.

GENUS *ANGULUS* MÜHLFELD, 1811

DIAGNOSIS: Shell small to large size, elongate-ovate, generally compressed, anterior end rounded, posterior end rostrate or subrostrate. There are two cardinals in each valve, right posterior and left anterior cardinals are bifid; right anterior and left posterior cardinals small and lamellar. Typically there is only a right anterior lateral adjacent to the cardinals, but in some species there may be an obsolete left anterior lateral corresponding to the one on the right valve. Posterior laterals are absent in both valves. Pallial sinus is of various size in different species, but the ventral margin of it is mostly or entirely coalescent with pallial line.

Key to the Subgenera of Genus *Angulus*

A. Sculpture concentric only.
    a. Sculpture of concentric striae. . . . .
        b. Pallial sinus large. . . . . *Angulus* s.s
        bb. Pallial sinus small. . . . . *Megangulus*
                                    new subgenus
    aa. Sculpture of concentric lamellae. . . . . *Tellinangulus*
B. Concentric and oblique sculpture. . . . . *Scissula*

Subgenus *Angulus* Mühlfeld

Angulus Mühlfeld, 1811, p. 47.

TYPE-SPECIES: *Tellina lanceolata* Gmelin (by subsequent designation Gray, 1847).

GEOLOGIC RANGE: Eocene to Recent.

Angulus (*Angulus*) *lanceolata* (Gmelin)
(Pl. 26, figs. 13–17)

Tellina lanceolata Gmelin, 1791, p. 3236; Wood, 1815, p. 167, Pl. 45, fig. 2; Hanley, 1846, p. 291, Pl. 58, figs. 86, 87; Römer, 1872, p. 125, Pl. 1, fig. 6, Pl. 28, figs. 11–13. *Psammotaea pellucida* Lamarck, 1818, p. 517.

DESCRIPTION: Shell of medium size, thin, elongate-ovate, compressed, inequilateral, inequivalve, with suffused rose coloring. The anterior end is rounded and slightly shorter than the posterior end which is rostrate, slightly gaping, and with acuminate extremity. Dorsal margins are sloping very gently, the postero-dorsal margin is slightly excavated behind the beak. Ventral margin is scarcely convex and its posterior end is emarginate. Postero-umbonal fold is of moderate strength on the right valve, but obsolete on the left valve. The ligament is small and situated in an external ligamental groove. Sculpture consists of fine concentric ridges which become obsolete on the posterior area of the left valve. The hinge is narrow and the teeth are weak; right posterior and left anterior cardinals are bifid, right anterior and left posterior cardinals small and lamellar. Right anterior lateral is adjacent to cardinals; the other laterals are absent. Inside of the shell there is a sculpture of very fine radial striae. Pallial sinus is large, the posterior half of its dorsal margin is slightly ascending and the anterior half descending steeply, its anterior end is acutely rounded and is only 2 millimeters from anterior adductor muscle scar. The ventral margin of pallial sinus is entirely coalescent with pallial line. Anterior adductor muscle scar is elongate and larger than the posterior one which is of orbicular shape.

FIGURED SPECIMEN: USNM 128454 (Malacca); length, 40.2 mm (100); height, 16.8 mm (41.8); thickness, 4.7 mm (11.7).

HABITAT: Malacca, Philippines and East Indies.

Subgenus *Megangulus*, New Subgenus

DIAGNOSIS: Shell of medium to large size, thick, white, elongate-ovate, compressed, anterior end rounded, posterior end sub-rostrate. The beaks are located slightly posterior to the mid-length; dorsal margins are sloping equally and steeply, ventral margin is broadly arcuate. Postero-umbonal fold is weak; ligament is external and of medium size. The sculpture consists of fine concentric striae. There are two *Tellina*-type cardinals in each valve, right anterior lateral is situated close to the cardinals and the other laterals are absent. Pallial sinus is small and ventral margin of it is entirely coalescent with the pallial line.

TYPE-SPECIES: *Tellina venulosa* Schrenck.

GEOLOGIC RANGE: Recent.

Angulus (Megangulus) venulosa (Schrenck)
Pl. 27, figs. 1–5

Tellina venulosa Schrenck, 1861, p. 412; Schrenck, 1867, p. 556, Pl. 22, figs. 2–5; Römer, 1872, p. 121, Pl. 28, figs. 8–10.

DESCRIPTION: Shell large, rather thick, somewhat porcelaneous, elongate-ovate, compressed, subequivalve, inequilateral. The anterior end is rounded; posterior end subrostrate with acuminate extremity and slightly bent to the right. Postero-umbonal fold is of moderate strength on the right valve, but it is obsolete on the left valve; ventral margin is broadly arcuate and slightly twisted. The beaks are situated slightly posterior to the mid-length; a medium size ligament is placed in an external ligamental groove. The color of the shell is white; there are some short, discontinuous, brown radial lines on the younger third of the shell, and a narrow strip along the entire margin of the shell is covered with periostracum. Sculpture consists of widely spaced growth lines and fine concentric striae; the former is well-marked on the younger portion of the shell, and the latter becomes slightly coarse on the posterior area. There are two cardinals in each valve, right posterior and left anterior cardinals are well-developed and bifid; right anterior and left posterior cardinals are small and lamellar. Right anterior lateral is well-developed and close to the cardinals; the other laterals are absent. The inside of the shell has a light orange color. Pallial sinus is small, its dorsal margin is straight, its anterior end rounded and 12 millimeters from the anterior adductor scar, the ventral margin of it is entirely coalescent with pallial line. The anterior adductor muscle scar is elongate and slightly larger than the posterior one, which has suborbicular shape.

FIGURED SPECIMEN: USNM 228958 (Otaru, Japan); length, 67 mm (100); height, 42.2 mm (63.0); thickness, 14.8 mm (22.1).

HABITAT: Northern Islands of Japan and Sakhalin.

Subgenus *Tellinangulus* Thiele

Tellinangulus Thiele, 1934, p. 919.

TYPE-SPECIES: *Angulus* (*T.*) *aethiopicus* (Jäeckel and Thiele) (= *Tellina* (*Tellinella*) *aethiopicus* Thiele and Jäeckel) (by monotypy).

GEOLOGIC RANGE: Recent.

Angulus (*Tellinangulus*) *aethiopicus* (Jäckel and Thiele)
(Pl. 28, fig. 1)

Tellina (*Tellinella*) *aethiopicus* Thiele and Jaeckel, 1931, p. 235, Pl. 4, fig. 110.

COMMENT: I had no access to a specimen of *A. aethiopicus*, and the following description is obtained from the observations made on the original illustration together with the brief diagnosis given by Thiele.

DESCRIPTION: Shell small, trigonal-ovate, anterior end rounded, posterior end short, rostrate and slightly bent to right. Umbones are almost centrally situated, dorsal margins slope steeply, ventral margin is convex and its posterior end is sinuous. Postero-umbonal fold is of moderate strength. Sculpture consists of small concentric lamellae with fine striae in the interspaces of the lamellae. Left anterior and right posterior cardinals are bifid; right anterior lateral are very short and adjacent to anterior cardinal.

The measurements of the shell are given as: length, 7.5 mm, height, 4.5 mm.

FIGURED SPECIMEN: I have copied the original figure given by Jaeckel and Thiele.

LOCALITY: East Africa.

Subgenus *Scissula* Dall

Scissula Dall, 1900, p. 291.

TYPE: *Tellina similis* Sowerby (by original designation).

GEOLOGIC RANGE: Lower Miocene to Recent.

Angulus (*Scissula*) *similis* (Sowerby)
(Pl. 28, figs. 2–6)

Tellina similis Sowerby, 1806, Pl. 75; Turton, 1819, p. 170; Hanley, 1846, p. 285, Pl. 57, fig. 65.

Tellina decora Say, 1827, p. 219.

Tellina iris Philippi, 1845, p. 25, Pl. 3, fig. 5 (not Say, 1822).

Tellina (*Angulus*) *decora* H. & A. Adams, 1856, p. 397.

DESCRIPTION: Shell small, rather thin, elongate-ovate, inequilateral, subequivalve, slightly compressed, white, often with rose-colored rays. The anterior end is longer and rounded; posterior end bluntly truncate, gaping, and slightly bent to right. The dorsal margins slope equally at a moderate angle; ventral margin is broadly arcuate. Ligament is small and external, postero-umbonal fold is obsolete. Sculpture on both valves consists of fine concentric growth lines on which are superimposed fine yet sharp incised rugae that are oblique on the middle area, but concentric on the anterior quarter of the shell and on the posterior area. Right posterior cardinal is relatively large, trigonal, and strongly bifid; right anterior cardinal is small and pointed. In the left valve the anterior cardinal is small and bifid, the posterior cardinal is lamellar, thin, and partly fused with the nymph. Right anterior lateral is strong and adjacent to cardinals; the other cardinals are absent. Pallial sinus is large, its dorsal margin slightly high in the middle, its anterior end is rounded and only one millimeter from anterior adductor muscle scar, the ventral margin of it is entirely coalescent with the pallial line. Anterior adductor muscle scar is elongate and slightly smaller than the posterior one which is of orbicular shape.

FIGURED SPECIMEN: USNM 95608 (Colon, Panama); length, 19.5 mm (100); height, 13.0 mm (66.7); thickness, 6.0 mm (30.8).

HABITAT: From South Carolina through the Caribbean Sea to Venezuela.

GENUS OMALA SCHUMACHER, 1817

Omala Schumacher 1817, p. 218; Salisbury, 1934 p. 77.

Homala 'Schumacher' Agassiz, 1846, p. 258 (not Mörch, 1852); Agassiz, 1848, p. 744; H. & A. Adams, 1856, p. 398; Fischer, 1887, p. 1148; Thiele, 1934, p. 919.

DIAGNOSIS: Shell small to large size, white, rather thin, compressed, elongate-oval, strongly inequilateral. Anterior end is short and broadly rounded, posterior end is slightly narrow and acutely rounded. Umbones are situated slightly less than a third of the length from anterior end, and beaks are low. Antero-dorsal margin is straight and has a gentle slope; ventral margin is broadly convex. A short and stout liagment is situated in a depressed ligamental groove, and postero-umbonal fold is moderate. Sculpture consists of fine concentric ridges. There are two cardinals in each valve; right laterals are small and adjacent to cardinals. Pallial sinus is small and its ventral margin is entirely coalescent with pallial line.

TYPE: *Tellina inaequivalvis* Schumacher (= *hyalina* Gmelin) (by monotypy).

GEOLOGIC RANGE: Recent.

Omala hyalina (Gmelin)
(Pl. 28, figs. 7–10)

Tellina hyalina Gmelin, 1791, p. 3235; Hanley, 1846, p. 292, Pl. 61, fig. 167; Reeve, 1867, Pl. 38, fig. 216; Römer, 1872, p. 181, Pl. 1, fig. 5.

Tellina complanata Röding, 1798, p. 187.

Omala inequivalvis Schumacher, 1817, p. 129, Pl. 10, fig. 1.

DESCRIPTION: Shell of large size, white, rather thin, compressed, elongate-oval, strongly inequilateral. Anterior end is strikingly short and broadly rounded, posterior end is slightly narrow and acutely rounded. Umbones are situated slightly less than a third of the length from anterior end, beaks are low and slightly opisthogyrate. Antero-dorsal margin slopes down with a convex curve, postero-dorsal margin is straight and has a gentle slope, and ventral margin is broadly arcuate. Postero-umbonal fold is moderate and more prominent on the right valve than it is on the left valve; ligament is short, stout and situated in a depressed ligamental groove. Sculpture consists of fine concentric ridges which are somewhat irregular throughout the shell surface except near the dorsal margins where they unite forming well-defined and slightly coarse ridges. There are some very fine radial striae which are better observable on both ends of the shell. Hinge plate is thin; on the right valve there are two thin cardinals and a lamellar anterior lateral adjacent to cardinals. Left anterior cardinal is thin and situated directly below the beak, left posterior cardinal is almost completely fused with the nymph and only small portion of it is observable with the aid of lens; left anterior lateral is obsolete. Posterior laterals of both valves are absent. In the interior of the valve there are fine radial ribs all of which converge toward the beak. Pallial sinus is small, its dorsal margin is steeply descending, its anterior end is acutely rounded and fourteen millimeters from anterior adductor muscle scar, and its ventral margin is completely coalescent with pallial line. Anterior adductor scar is elongate and larger than the posterior adductor scar; the latter has a bend on its upper part.

FIGURED SPECIMENS: USNM 128651 (Liberia, West Africa); length, 57.4 mm (100); height, 34.2 mm (59.6); thickness (one valve), 5.5 mm (19.2); small valve; USNM 123467 (Liberia, West Africa).

HABITAT: West Africa, from Guinea to Angola.

GENUS PHYLLODA SCHUMACHER, 1817

Phylloda Schumacher, 1817, pp. 148, 149.

DIAGNOSIS: Shell large to medium size, moderately thin, compressed, elongate-ovate, subequilateral, subequivalve, glossy, orange colored. Anterior end is short and rounded, posterior end is obliquely truncate. Dorsal margins slope gently, postero-dorsal margin is spinous, ventral margin is slightly arcuate, and postero-umbonal fold is moderate. Sculpture consists of fine concentric ridges, and small scales on the posterior area. There are two cardinals and an anterior lateral in each valve. Pallial sinus is rather small and more than half of its ventral margin is free.

TYPE: *Tellina foliacea* Linnaeus (by monotypy).

GEOLOGIC RANGE: Tertiary to Recent.

Phylloda foliacea (Linnaeus)
(Pl. 29, figs. 1–3; Pl. 30, figs. 1–2)

Tellina foliacea Linnaeus, 1758, p. 675; Römer, 1871, p. 162, Pl. 3, fig. 13, Pl. 34, figs. 1–3.

Phylloda aurea Schumacher, 1817, p. 149, Pl. 16, figs. 1.

Phylloda foliacea Mörch, 1852, p. 11, no. 100.

DESCRIPTION: Shell large, rather thin, orange colored, glossy, elongate-ovate, compressed, subequilateral, subequivalve. Anterior end is rounded, attenuated, and somewhat narrow; posterior end is obliquely truncate, and slightly broader than the anterior end. Beaks are low and situated slightly anterior to mid-length, the umbones are greatly subdued, and postero-umbonal fold is moderate. Antero-dorsal margin has a gentle slope, postero-dorsal margin is convex and spinous; ventral margin is only moderately arcuate. Ligament is lanceolate, slender, and long, extending along the entire length of postero-dorsal margin and situated in depressed ligamental groove. Sculpture consists of fine concentric ridges and very fine striae; the posterior area is studded with rows of small scales which are more numerous on the posterior area of right valve than on the left. Hinge plate is thin, and there are two cardinals with an anterior lateral in each valve. On the right valve, anterior cardinal is small, the posterior one is large, trigonal and bifid; right anterior lateral is small and adjacent to cardinals. On the left valve, the anterior cardinal is small and bifid, the posterior one is long and slender; left anterior lateral is obsolete. Posterior laterals of both valves are absent. Pallial sinus is small, narrow, and ascending, its anterior end extending only to the middle of the shell, and slightly more than half of its ventral margin is free; there is a line joining anterior end of pallial sinus with anterior adductor muscle scar. Anterior adductor muscle scar is elongate, the posterior one is suborbicular.

FIGURED SPECIMEN: USNM 217195 (Red Sea); length, 66.4 mm (100); height, 35.8 mm (54.0); thickness, 18.0 mm (27.2).

HABITAT: Red Sea, Indian Ocean, to Philippines.

GENUS BARYTELLINA MARWICK, 1924

THE ORIGINAL DIAGNOSIS: "Shell ovate, thick, flexuous, surface with irregular, weak, concentric threads, right valve with a very large, thick, grooved posterior cardinal tooth, and a lamellar anterior one, anterior lateral obsolete or quite absent, posterior lateral very strong, left valve with posterior cardinal lamellar, anterior thick and grooved, no anterior lateral, posterior lateral distant and strong; sinus coalescing with the pallial line right from the anterior adductor" (Marwick).

TYPE: *Barytellina crassidens* Marwick (by original designation).

Subgenus Barytellina Marwick

Barytellina Marwick, 1924, p. 25.

TYPE: *Barytellina crassidens* Marwick.

GEOLOGIC RANGE: Pliocene.

Barytellina (Barytellina) crassidens Marwick
(Pl. 30, figs. 3–7)

Barytellina crassidens Marwick, 1924, p. 26, figs. 3, 5.

COMMENT: I was unable to obtain a specimen of this species, therefore I am quoting the original description which is as follows:

DESCRIPTION: "Shell ovate, thick, only slightly flexuous; sculpture consisting of fine, irregular, concentric ridges; a strong radial ridge runs from the umbo to the posterior end, with another narrow ridge bounding its posterior side; teeth as above; there is a slight widening of the hinge margin some distance in front of the anterior cardinal, that may represent an atrophied anterior lateral; anterior adductor long and narrow, posterior ovate, pallial sinus undulating, coalescing with the pallial line; there are indistinct internal radials above the sinus, two close together and fairly distinct above the posterior adductor."

"Holotype in the collection of the New Zealand Geological Survey."

"Dimensions: length 29, height 21, diameter (one valve) 7 mm."

FIGURED SPECIMEN: The original figure given by Marwick has been copied.

LOCALITY: Nukumaru Beach, Wanganui, New Zealand.

GENUS MACOMA LEACH, 1819

DIAGNOSIS: Shell large to very small size, white, rather thin, subtrigonal-ovate, inequilateral. Anterior end is rounded, posterior end is bluntly truncate; ligament is somewhat small, postero-umbonal fold is obsolete, and sculpture is concentric or oblique. There are two cardinals in each valve, right posterior and left anterior cardinals are bifid; laterals are absent. Pallial sinus is of large or medium size and partly coalescent with pallial line below.

TYPE: *Tellina calcarea* Gmelin, 1791 (by monotypy).

Key to the Subgenera of Genus Macoma

A. Beaks situated posterior to mid-length.
    a. Anterior end considerably longer than posterior end.
        b. Shell larger than 12 mm.
            c. Pallial sinus discrepant in two valves.
                (1) Pallial sinus not coalescent with anterior adductor . . *Macoma* s.s.
                (2) Pallial sinus coalescent with anterior adductor. . . . *Austromacoma*
            cc. Pallial sinus same in two valves.
                d. Shell elongate-ovate.
                    e. Sculpture of concentric striae only.

      f. Shell larger than 32 mm.
         (1) Shell inflated. . . . . *Psammacoma*
         (2) Shell compressed. . . . . *Ardeamya*
      ff. Shell smaller than 32 mm.
         g. Dorsal margin of pallial sinus smooth. . . . . *Macomopsis*
         gg. Dorsal margin of pallial sinus undulating. . . *Cydippina*
    ee. Sculpture of concentric striae and scattered granu-
    lations.
         (1) Cardinal teeth small . . . . *Macoploma*
         (2) Cardinal teeth strong and tending to coalesce
            with cardinal fulcrum. . . . . *Bendemecoma*
  dd. Shell trigonal. . . . . *Pseudometis*
bb. Shell smaller than 12 mm.
  h. Shell trigonal. . . . . *Pinguimacoma*
  hh. Shell elongate-ovate.
    i. Sculpture concentric only. . . . . *Exotica*
    ii. Sculpture concentric and oblique. . . . . *Loxoglypta*
aa. Anterior end only slightly longer than posterior end.
  j. Posterior end bluntly truncate.
    k. Sculpture concentric.
      l. Shell larger than 45 mm.
        m. Shell orbicular. . . . . *Rexithaerus*
        mm. Shell elongate-ovate. . . . . *Psammotreta*
      ll. Shell smaller than 45 mm. . . . . *Salmacoma*
    kk. Sculpture oblique.
      n. Shell larger than 12 mm.
        o. Sculpture oblique on both valves.
          (1) Sculpture of incised rugae covering entire shell
            surface. . . . . *Temnoconcha*
          (2) Sculpture of incised rugae covering any anterior
            part of shell surface. . . . . *Psammothalia*
        oo. Sculpture oblique on right valve and concentric on
          left valve. . . . . *Scissulina*
      nn. Shell smaller than 12 mm.
        p. Sculpture of concentric and oblique striae.. . . . . *Jactellina*
        pp. Sculpture of concentric striae only. . . . . *Ascitellina*
  jj. Posterior end rostrate.
    q. Sculpture oblique.. . . . . *Cymatoica*
    qq. Sculpture concentric. . . . . *Bartrumia*
B. Beaks situated anterior to mid-length.
  a. Sculpture concentric only.
    b. Sculpture of fine concentric ridges.
      c. Dorsal margin of pallial sinus straight. . . . . *Peronidia*
      cc. Dorsal margin of pallial sinus convex. . . . . *Bartschicoma*
                             new subgenus
    bb. Sculpture of concentric lamellae. . . . . *Rostrimacoma*
  aa. Sculpture of concentric striae and small granulations. . . . . *Panacoma*

Subgenus *Macoma* Leach

Macoma Leach, 1819, p. 62.
Macoma Gray, 1825, p. 136.

Limicola Leach, 1852, p. 296 (not *Limicola* Koch, Aves, 1816, nor Fischer, 1887).
TYPE: *Macoma tenera* Leach (= *calcarea* Gmelin) (by monotypy).
GEOLOGIC RANGE: Eocene to Recent.

Macoma (Macoma) calcarea (Gmelin)
(Pl. 30, figs. 8–12)

Tellina calcarea Gmelin, 1791, p. 3236; Römer, 1872, p. 222, Pl. 34, figs. 1–6.
Tellina lata Gmelin, 1791, p. 3237.
Tellina sabulosa Spengler, 1798, p. 114.
Macoma tenera Leach, 1819, p. 62.
Tellina sordida Couthouy, 1838, p. 59, Pl. 3, fig. 11.
Tellina proxima Sowerby, 1839, p. 154, Pl. 44, fig. 4; Hanley, 1846, p. 313, Pl. 66, fig. 264, Pl. 59, fig. 115.
Sanguinolaria sordida Gould, 1841, p. 67.
GEOLOGIC RANGE: Recent.

DESCRIPTION: Shell of medium size, rather thin, subtrigonally ovate, somewhat compressed, inequilateral, white usually with brown periostracum. Anterior end is longer than the posterior end and broadly rounded, posterior end is narrow, bluntly truncate, and slightly bent to right. Beaks are low and situated about 0.35 of the length from posterior end; antero-dorsal margin has gentle slope, postero-dorsal margin is steep; ventral margin is broadly convex. Ligament is lanceolate, rather small, and situated in a shallow ligamental groove; postero-umbonal fold is obsolete, and sculpture consists of fine concentric striae. Hinge plate is thin, and there are two weak cardinals in each valve; right anterior and left posterior cardinals are small and thin, right posterior and left anterior cardinals are bifid and slightly larger than the other cardinals; laterals are absent in both valves. Pallial sinus is discrepant in two valves; that of right valve is small, its dorsal margin is anteriorly descending, the anterior end of it is acutely rounded and extends forward to within 9 millimeters from anterior adductor muscle scar, and posterior half of its ventral margin is coalescent with pallial line. The pallial sinus of left valve is large, its dorsal margin is nearly horizontal, the anterior end of it is broadly rounded and extends forward to within 3 millimeters of anterior adductor muscle scar, and posterior half of its ventral margin is coalescent with pallial line. Anterior adductor muscle scar is slightly larger than the posterior one and has elongate shape, posterior one is suborbicular.

FIGURED SPECIMEN: USNM 338474 (Lunenburg, Nova Scotia); length, 41.4 mm (100); height, 29.1 mm (70.3); thickness, 12.0 mm (29.0).

HABITAT: Members of the species *M. calcarea* occur at the present time in cooler seas, especially in arctic and subarctic waters, and extend southward as far as Long Island Sound on the Atlantic Coast and to Oregon on the Pacific Coast of North America. They are also found along the high-latitude coastal waters of both Europe and Asia.

Subgenus Rexithaerus Tryon

Rexithaerus Tryon, 1869, p. 104.
TYPE: *Macoma secta* (Conrad) (by subsequent designation Dall, 1900).
GEOLOGIC RANGE: Upper Oligocene to Recent.

Macoma (Rexithaerus) secta (Conrad)
(Pl. 31, figs. 1–5)

Tellina secta Conrad, 1837, p. 257; Hanley, 1846, p. 327, Pl. 65, figs. 245, 248; Römer, 1872, p. 260, Pl. 50, figs. 1–5.

DESCRIPTION: Shell large, white, rather thin, trigonally-ovate, somewhat compressed, subequilateral. Anterior end is slightly longer than the posterior end and broadly rounded, posterior end is bluntly truncate and gaping. Dorsal margins are steep and slope down with an equal angle on both sides, postero-dorsal margin is excavated behind the beak; ventral margin is convex. Postero-umbonal fold is moderately strong on the right valve and weak on the left valve, ligament is short but stout and situated in a shallow ligamental groove. Sculpture consists of fine concentric striae, and a narrow zone along the margins of the shell is covered with periostracum. Right posterior and left anterior cardinals are large, bifid, and with a deep socket on each side; right anterior and left posterior cardinals are small and thin, laterals of both valves are absent. Pallial sinus is discrepant in the two valves, that of right valve being smaller than the one in left valve; in both valves the dorsal margin of pallial sinus is high posteriorly and descending anteriorly, its anterior end is acutely rounded, and almost the entire ventral margin of it is coalescent with pallial line; but in the right valve the anterior end of pallial sinus is separated from the anterior adductor muscle scar by a distance of 9 millimeters. Anterior adductor muscle scar is elongate, the posterior one is suborbicular.

FIGURED SPECIMEN: USNM 172954 (Bolinas, California); length, 64.4 mm (100); height, 52.8 mm (82.0); thickness, 20.1 mm (31.2).

HABITAT: Puget Sound to Baja California, Mexico.

Subgenus *Psammacoma* Dall

Psammacoma Dall, 1900, p. 292.

TYPE: *Psammotaea candida* (Lamarck) (by original designation).

GEOLOGIC RANGE: Recent.

Macoma (Psammacoma) candida (Lamarck)
(Pl. 32, figs. 1–5)

Psammotaea candida Lamarck, 1818, p. 517.
Tellina galatea Römer, 1872, p. 249, Pl. 47, figs. 7–8.
Macoma candida Bertin, 1878, p. 342.

DESCRIPTION: Shell moderately large, white, rather thin, elongate-ovate, strongly inequilateral, subequivalve, inflated, anterior end rounded, posterior end bluntly truncate. Umbones are situated one third of the length from posterior end, antero-dorsal margin has gentle slope, postero-dorsal margin slopes down steeply, ventral margin is straight. Ligament is short and situated in a slightly depressed ligamental groove, postero-umbonal fold is obsolete. The sculpture consists of fine concentric and very fine radial striae. Hinge plate is thin, and there are two cardinals in each valve; in the right valve the anterior cardinal is small and posterior one is large, bifid, with a deep socket on each side; cardinals of left valve are small and the anterior one is bifid; laterals of both valves are absent. Pallial sinus is of medium size, its dorsal margin is high and nearly straight, its anterior end is rounded and separated from anterior adductor muscle scar by a distance of 12 millimeters, posterior half of its ventral margin is coalescent with pallial line. Anterior adductor muscle scar is larger than posterior one and is elongate, the posterior one is suborbicular.

FIGURED SPECIMEN: USNM 47936 (You ma ti, Hong Kong); length, 57.6 mm (100); height, 33.6 mm (58.4); thickness, 16.9 mm (29.4).

HABITAT: East Indies to Japan and Australia.

Subgenus *Macoploma* Pilsbry and Olsson

Macoploma Pilsbry and Olsson, 1941, p. 69.
TYPE: *Macoma ecuadoriana* Pilsbry and Olsson (by original designation).
GEOLOGIC RANGE: Pliocene.

Macoma (Macoploma) ecuadoriana Pilsbry and Olsson
(Pl. 32, fig. 6)

Macoma ecuadoriana Pilsbry and Olsson, 1941, p. 69, Pl. 19, fig. 5.
COMMENT: I was unable to obtain a specimen of *M. ecuadoriana*, therefore I am quoting the original description:
DESCRIPTION: "Shell elongate, the valves subequal but with the left a little larger and more convex than the right, which is somewhat depressed or flexed across the middle; beaks situated a little in front of the posterior third, small, pointed and adjacent; anterior side long, the posterior short, truncated at its end; surface marked with lines of growth and a scattered granulation, best developed on the posterior submargins, fine or absent elsewhere; the hinge of the left valve has a small, partly double cardinal tooth but no laterals."
FIGURED SPECIMEN: ANSP 13709 (Ecuador); length, 61.0 mm (100); height, 33.0 mm (54.0); thickness, 16.5 mm (27.0).
OCCURRENCE: Canoa formation, Punta Blanca, Ecuador.
REMARKS: Pilsbry mentions that this shell resembles *Macoma lamproleuca* Pilsbry & Lowe, but is distinguished by the coarse granulation of its posterior area.

Subgenus *Macomopsis* Sacco

Macomopsis Sacco, 1901, p. 107.
TYPE: *Tellina elliptica* Brocchi (by original designation).
GEOLOGIC RANGE: Pliocene.

Macoma (Macomopsis) elliptica (Brocchi)
(Pl. 32, figs. 7–10)

Tellina elliptica Brocchi, 1814, p. 513, Pl. 12, figs. 7a, b.
DESCRIPTION: Shell small, white, rather thin, elongate-ovate, strongly inequilateral, moderately inflated, subequivalve, anterior end rounded, posterior end bluntly truncate. Umbones are situated one third of the length from posterior end, antero-dorsal margin has gentle slope, postero-dorsal margin slopes down steeply, ventral margin is straight. Escutcheon is short and narrow, postero-umbonal fold is obsolete, and the sculpture consists of fine concentric striae. Hinge plate is thin and there are two cardinals in each valve; in the right valve the anterior cardinal is small, posterior one is large, trigonal, and bifid; left anterior cardinal is bifid and smaller than the bifid cardinal of right valve, left posterior cardinal is small and thin; laterals of both valves are absent. Pallial sinus is of medium size, its dorsal margin is only slightly descending anteriorly, its anterior end is rounded and is separated from anterior adductor muscle scar by a distance of 6 millimeters, posterior half of its ventral margin is coalescent with pallial line. Anterior adductor muscle scar is slightly larger than posterior one and elongate in shape, the posterior one is suborbicular.
FIGURED SPECIMEN: USNM 201862 (Mt. Pellegrino, Palermo, Sicily); length, 31.4 mm (100); height, 19.6 mm (62.4); thickness, 5.6 mm (17.8).
OCCURRENCE: Mt. Pellegrino, Sicily, and Modena, Italy.

Subgenus *Cydippina* Dall

Cydippina Dall, 1900a, p. 292.

TYPE: *Macoma brevifrons* Say (by original designation).

GEOLOGIC RANGE: Pliocene to Recent.

Macoma (*Cydippina*) *brevifrons* Say
(Pl. 33, figs. 1–5)

Tellina brevifrons Say, 1830, Pl. 64, fig. 1; Tryon, 1874, Pl. 26, figs. 355–357.

DESCRIPTION: Shell small, thin, white with orange coloring on the umbonal area, elongate-ovate, subequivalve, strongly inequilateral, anterior end rounded, posterior end bluntly truncate. Umbones are situated one-third of the length from posterior end, antero-dorsal margin has gentle slope, postero-dorsal margin slopes down steeply, ventral margin is nearly straight. The ligament is small, lanceolate, and situated in a depressed ligamental groove; postero-umbonal fold is obsolete, and the sculpture consists of fine concentric striae. Hinge plate is thin, there are two cardinals in each valve; right anterior and left posterior cardinals are small and thin, right posterior and left anterior cardinals are bifid and larger than the other cardinals; laterals of both valves are absent. Pallial sinus is large, its dorsal margin is posteriorly high and slightly descending anteriorly, its anterior end is rounded and separated from anterior adductor muscle scar by a space of three millimeters, posterior half of its ventral margin is coalescent with pallial line. The muscle scars are subequal, anterior one is elongate, and the posterior one is orbicular.

FIGURED SPECIMEN: USNM 17687 (West Indies); length, 23.0 mm (100); height, 13.9 mm (60.5); thickness, 6.5 mm (28.3).

HABITAT: South Carolina to Brazil.

REMARKS: Individuals of this species are small toward the northern extreme of its range, but attain considerable size (length, 39; height, 23.5, and thickness, 13 mm) in the warmer waters to the south and in those of Pliocene times.

Subgenus *Pseudometis* Lamy

Pseudometis Lamy, 1918, p. 170.

TYPE-SPECIES: *Tellina truncata* Jonas (= Tellina praerupta Salisbury) (by subsequent
designation Salisbury, 1929).

GEOLOGIC RANGE: Recent.

Macoma (*Pseudometis*) *praerupta* Salisbury
(Pl. 33, figs. 6–10)

Tellina truncata Jonas, in Philippi, 1844, p. 71, Pl. 1, fig. 2 (not Linnaeus, 1767); Hanley, 1846, p. 325, Pl. 62, fig. 198; Reeve, 1866, Pl. 8, sp. 33; Römer, 1872, p. 248, Pl. 47, figs. 4–6.

Tellina praerupta Salisbury, 1934, p. 90.

DESCRIPTION: Shell of medium size, white, somewhat thin, trigonal-donaciform, rather compressed, nearly equivalve, strongly inequilateral. Anterior end is narrower than the posterior end and sharply rounded, posterior end is bluntly truncate. Umbones are situated one-third of the length from posterior end, postero-dorsal margin is sloping very steeply, the antero-dorsal margin slopes gently; ventral margin is only slightly arcuate. The ligament is small and situated in depressed ligamental groove, postero-umbonal fold is obsolete. Sculpture consists of fine concentric ridges which become obsolete on the umbonal area. Hinge plate is thin; right valve has two well-developed cardinals, the anterior one is small; posterior one is trigonal, bifid, and large. In the left valve, the anterior cardinal is trigonal, large, and bifid; the posterior one is small, thin, and partly fused with the nymph; laterals of both valves are absent. Pallial sinus is of medium size, its dorsal margin is high in the middle, anterior end of it is broadly rounded and separated from anterior adductor muscle scar by a distance of nine millimeters, and posterior half of its ventral margin is coalescent with pallial line. Anterior adductor muscle scar is larger than the posterior one and is elongate in shape, the posterior one is suborbicular.

FIGURED SPECIMEN: USNM 17883 (China Sea); length, 39.8 mm (100); height, 29.6 mm (75.0); thickness, 13.4 mm (33.7).

HABITAT: Red Sea and East Africa to Philippines and Japan.

Subgenus Psammotreta Dall

Psammotreta Dall, 1900, p. 292.

TYPE: *Tellina aurora* Hanley (by original designation).

GEOLOGIC RANGE: Recent.

Macoma (Psammotreta) aurora (Hanley)
(Pl. 34, figs. 1–5)

Tellina aurora Hanley, 1844, p. 147; Hanley, 1846, p. 301, Pl. 58, fig. 76; Römer, 1872, p. 244, Pl. 46, figs. 7–9.

DESCRIPTION: Shell of medium size, somewhat thin, white, elongate-ovate, slightly compressed, nearly equivalve, subequilateral. Anterior end is slightly longer than posterior end and broadly rounded, posterior end is bluntly truncate. Umbones are situated slightly posterior of mid-length, antero-dorsal margin has gentle slope, postero-dorsal margin slopes down steeply; ventral margin is broadly arcuate. Ligament is small and situated in a small ligamental groove, postero-umbonal fold is obsolete. The sculpture consists of fine concentric striae and some widely spaced growth lines. Right anterior and left posterior cardinals are small and lamellar, right posterior and left anterior cardinals are bifid and slightly larger than the other cardinals; laterals of left valve are absent. Pallial sinus is of medium size, the dorsal margin of it is posteriorly high and anteriorly descending, its anterior end is rounded and separated from anterior adductor muscle scar by a distance of seven millimeters, and posterior two-thirds of its ventral margin is coalescent with pallial line. Adductor muscle scars are subequal, the anterior one is elongate, and posterior one suborbicular.

FIGURED SPECIMEN: USNM 120663 (Guaymas, Mexico); length, 45.0 mm (100); height, 29.0 mm (64.5); thickness, 11.2 mm (24.9).

HABITAT: Lower California to Ecuador.

Subgenus Temnoconcha Dall

Temnoconcha Dall, 1921, p. 132.

TYPE-SPECIES: *Macoma brasiliana* Dall (by monotypy).

GEOLOGIC RANGE: Recent.

Macoma (Temnoconcha) brasiliana (Dall)
(Pl. 34, figs. 6–10)

Psammacoma (Temnoconcha) brasiliana Dall, 1921, p. 132.

DESCRIPTION: Shell of medium size, white, somewhat thin, compressed, equivalve, subequilateral. The anterior end is slightly longer than the posterior end and broadly rounded, posterior end is bluntly truncate and slightly gaping. Umbones are situated slightly posterior of the mid-length, dorsal margins are sloping gently with equal angle on both sides and postero-dorsal margin is slightly excavated, ventral margin is broadly arcuate. The ligament is small, lanceolate, and situated in a depressed ligamental groove; postero-umbonal fold is obsolete. Sculpture consists of fine incised rugae which are concentric on the anterior quarter of the shell and become oblique over the remaining area of the shell up to posterior fold; these oblique rugae cross over rather widely spaced growth lines, which become crowded together on the posterior area and constitute the only sculpture on that area. Right valve has two bifid cardinals which are separated from each other by a deep socket; in the left valve the anterior cardinal is bifid, posterior is lamellar and partly fused with the nymph; laterals of both valves are absent. Pallial sinus is large, dorsal margin of it is nearly straight, its anterior end is broadly rounded and separated from anterior adductor muscle scar by two millimeters, posterior half of its ventral margin is coalescent with pallial line.

FIGURED SPECIMEN: USNM 333023 (San Sebastian Island, Brazil); length, 35.0 mm (100); height, 23.2 mm (66.3); thickness, 8.0 mm (22.8).

HABITAT: Southern coast of Brazil.

Subgenus *Scissulina* Dall

Scissulina Dall, 1924, p. 88.

TYPE-SPECIES: *Tellina dispar* Conrad (by original designation).

GEOLOGIC RANGE: Recent.

Macoma (Scissulina) dispar (Conrad)
(Pl. 34, figs. 11–15)

Tellina dispar Conrad, 1837, p. 259; Hanley, 1846, p. 306, Pl. 59, figs. 113, 114; Reeve, 1866, Pl. 3, figs. 10a, b, c; Römer, 1871, p. 148, Pl. 32, figs. 7–9.

DESCRIPTION: Shell small, white, rather thin, elongate-ovate, somewhat compressed, inequivalve, inequilateral. Anterior end is longer than the posterior end and smoothly rounded, posterior end is bluntly truncate and slightly gaping. Umbones are situated a short distance posterior of mid-length, dorsal margins are sloping equally at a moderate angle on both sides; ventral margin is broadly arcuate. Ligament is small and situated in a shallow ligamental groove, postero-umbonal fold is obsolete. The sculpture is discrepant on two valves, that of right valve consisting of fine incised rugae which are concentric on the anterior third of valve and become oblique over the remainder of the shell up to posterior fold beyond which they become obsolete; sculpture on the left valve consists of fine concentric striae and widely spaced very fine radial lines. Hinge plate is thin; right valve has two subequal cardinals, the posterior one is bifid and slightly larger than the other one; in the left valve the anterior cardinal is bifid and the posterior one is lamellar and mostly fused with the nymph. Laterals of both valves are absent. Pallial sinus is large, the dorsal margin of it is posteriorly high and anteriorly descending until it touches the pallial line one millimeter behind the anterior adductor muscle scar, ventral margin of it is entirely coalescent with pallial line. Adductor muscle scars subequal and both are suborbicular.

FIGURED SPECIMEN: USNM 33735 (Hilo, Hawaii); length, 28.0 mm (100); height, 18.3 mm (65.3); thickness, 8.0 mm (28.6).

HABITAT: Pacific and Indian oceans; reported from Hawaiian Islands, New Caledonia, Philippines, Wallis Islands, and Mauritius.

Subgenus *Peronidia* Dall

Peronidia Dall, 1900, p. 291.

TYPE-SPECIES: *Tellina albicans* Gmelin (by original designation).

GEOLOGIC RANGE: Pleistocene to Recent. Fossils of this species occur in the youngest shale formations of Sicily and Italy.

Macoma (Peronidia) albicans (Gmelin)
(Pl. 35, figs. 1–5)

Tellina albicans Gmelin, 1791, p. 3238.

Tellina nitida Poli, 1791, Pl. 15, figs. 2–4; Lamarck, 1818, p. 527; Hanley, 1846, p. 308, Pl. 59, fig. 101; Reeve, 1866, Pl. 13, fig. 57; Römer, 1872, p. 118, Pl. 3, fig. 12. Pl. 32, figs. 11–14.

DESCRIPTION: Shell of medium size, rather thin, white, trigonal-ovate, compressed, subequivalve, inequilateral. The anterior end is rounded and slightly shorter than posterior end which is sub-rostrate and bluntly truncate. Dorsal margins are sloping equally at a moderate angle; the postero-dorsal margin is slightly excavated behind the beak, ventral margin is broadly arcuate. A small lanceolate ligament is situated in a depressed ligamental groove; the postero-umbonal fold is obsolete. Sculpture consists of fine concentric ridges which become slightly oblique on the posterior two-thirds of the shell, and fine radial striae on the posterior half of the shell. Right valve has two bifid cardinals, the posterior one is larger than the anterior one; left anterior cardinal is bifid, the posterior cardinal of the left valve is thin, lamellar and mostly fused with the nymph. Laterals of both valves are absent. Pallial sinus is large, its dorsal margin nearly straight, the anterior end rounded and 4 millimeters from the anterior adductor muscle scar, and ventral margin of it is entirely coalescent with pallial line. Anterior adductor muscle scar is elongate and slightly larger than the posterior one which is orbicular in shape.

FIGURED SPECIMEN: USNM 304435 (Viareggio, Italy); length, 43.4 mm (100); height, 22.7 mm (52.3); thickness, 7.0 mm (16.1).

HABITAT: In the Mediterranean this species is reported from coast of France, Italy, Sicily, Corsica and Algiers.

Subgenus Bartschicoma, New Subgenus

DIAGNOSIS: Shell of moderately large size, white, trigonal-ovate, compressed, sub-equivalve, inequilateral. The anterior end is rounded, posterior end sub-rostrate and longer; sculpture consisting of fine concentric ridges. There are two cardinals in each valve, the laterals of both valves are absent. Pallial sinus is large, dorsal margin of it convex and ventral margin entirely coalescent with pallial line.

TYPE-SPECIES: *Tellina gaimardi* Iredale.

GEOLOGIC RANGE: Recent.

Macoma (Bartschicoma) gaimardi (Iredale)
(Pl. 35, figs. 6–10)

Tellina alba Quoy and Gaimard, 1835, p. 500, Pl. 80, figs. 1–3 (not Wood, 1815); Hanley, 1846, p. 113, Pl. 62, fig. 193; Reeve, 1867, Pl. 32, fig. 180; Römer, 1872, p. 255, Pl. 48, figs. 10–12.

Tellina gaimardi Iredale, 1915, p. 489.

DESCRIPTION: Shell moderately large, white, compressed, trigonal-ovate, subequi-valve. The anterior end is broadly rounded and slightly emarginate in the place where a feeble fold extends from below the beak and terminates at the lower end of anterior margin. The posterior end is sub-rostrate and long; the beaks are located slightly anterior to the mid-length. Dorsal margins are sloping at a moderately steep angle, the postero-dorsal margin is slightly excavated behind the beak; ventral margin is broadly arcuate. Postero-umbonal fold is obsolete, and a medium size ligament is situated in a depressed ligamental groove. Sculpture consists of fine concentric ridges which become somewhat coarse along the margins. Right valve has two cardinals, the posterior one is bifid and larger than the other; in the left valve the anterior cardinal is bifid, posterior one thin, lamellar, and partly fused with the nymph. Laterals of both valves are absent. Pallial sinus is large, its dorsal margin is high in the middle and descending anteriorly, the anterior end of it is within 2 millimeters of anterior adductor scar and ventral margin of it entirely coalescent with the pallial line. Anterior adductor muscle scar is elongate and slightly larger than the posterior one which is orbicular in shape.

FIGURED SPECIMEN: USNM 22886 (New Zealand); length, 75.0 mm (100); height, 39.4 mm (52.5); thickness, 11.4 mm (15.2).

HABITAT: New Zealand.

Subgenus Rostrimacoma Salisbury

Rostrimacoma Salisbury, 1934, pp. 78, 82.

TYPE-SPECIES: *Panopea cancellata* Sowerby (by original designation).

GEOLOGIC RANGE: Recent.

Macoma (Rostrimacoma) cancellata (Sowerby)
(Pl. 36, fig. 1)

Panopea cancellata Sowerby, II, 1873, Pl. 4, fig. 4.

COMMENT: I was unable to obtain a specimen of *P. cancellata;* therefore, I am quoting the original description given by Sowerby:

DESCRIPTION: "Shell ovate-oblong, fulvous, rather thin, a little gaping at both ends: posterior side smooth, sub-rostrate, depressed, slightly arched, with the dorsal margin depressed, concave, end narrow, rounded, ventral margin sloped upwards; middle convex, ventral margin rather straight; anterior side inflated, cancellated with concentric wrinkles and radiating striae, rounded at the end."

FIGURED SPECIMEN: Copied from Sowerby's original figure.

HABITAT: Australia.

Subgenus Panacoma Olsson

Panacoma Olsson, 1942, p. 195.

TYPE-SPECIES: *Macoma (Panacoma) chiriquiensis* Olsson (by original designation).

GEOLOGIC RANGE: Pliocene.

Macoma (Panacoma) chiriquiensis Olsson
(Pl. 36, figs. 2–3)

Macoma (Panacoma) chiriquiensis Olsson, 1942, p. 195, Pl. 5, figs. 5, 6.

COMMENT: I could not obtain a specimen of *M. chiriquiensis*; therefore, I am quoting the original description.

DESCRIPTION: "Shell thin, broadly sub-ovate, subequal; the anterior side is somewhat shorter and more widely rounded than the posterior; there is no posterior flexure so that the two sides appear more or less subequal; the sculpture is formed by fairly regular, small, raised threads which are weakly flexed in the middle portion of the ventral side; in addition, the surface has a sprinkling of small granules recalling some species of *Thracia;* body cavity deep with a wide pallial sinus reaching a little beyond the middle, below, it is confluent with the pallial line which reaches nearly to the posterior muscle scar; hinge of the left valve with normal *Macoma*-tooth formula, 2 cardinal teeth, the anterior one larger, no laterals."

FIGURED SPECIMEN: Holotype, Paleontological Research Institution, no. 5002; length, 34 mm; height, 27 mm; (estimated) semi-diameter, 7 mm.

OCCURRENCE: Charco Azul formation of Burica Peninsula of Panama and Costa Rica.

Subgenus Pinguimacoma Iredale

Pinguimacoma Iredale, 1936, p. 282.

TYPE-SPECIES: *Pinguimacoma hemicilla* Iredale (by original designation).

GEOLOGIC RANGE: Recent.

Macoma (Pinguimacoma) hemicilla Iredale
(Pl. 36, figs. 4–5)

Pinguimacoma hemicilla Iredale, 1936, p. 282, Pl. 21, fig. 7.

COMMENT: A specimen of *M. hemicilla* was not available; therefore, I am quoting the original description.

DESCRIPTION: "Shell small, thin, pinkish-white, inequilateral; a little swollen, smooth, growth lines showing only toward the ventral margin. The short beak is also smooth, and the hinge shows only cardinal teeth; the pallial line appears to agree with that of *Pinguitellina*, but is difficult to observe owing to the thinness of the shell. This is another of the forms which suggest that the loss of lateral teeth is recent, the hinge ligament probably compensating for this loss. From the species of *Pinguitellina* this little shell is separable externally by the short beak and more swollen anterior portion."

Length, 11 mm; height, 9 mm.

LOCALITY: Sydney Harbour.

FIGURED SPECIMEN: The original figure has been reproduced.

HABITAT: New South Wales.

Subgenus Jactellina Iredale

Jactellina Iredale, 1929, p. 266.

TYPE-SPECIES: *Tellina obliquaria* Deshayes (by original designation).

GEOLOGIC RANGE: Recent.

Macoma (Jactellina) obliquaria (Deshayes)
(Pl. 36, figs. 6–10)

Tellina obliquaria Deshayes, 1854, p. 356; Reeve, 1868, Pl. 54, fig. 321.

DESCRIPTION: Shell small, thin, white or white with rose coloring on the umbonal area, trigonal-orbicular, slightly inflated, subequivalve, inequilateral. Anterior end is longer than posterior end and rounded, the posterior end is obliquely truncate and somewhat wider than anterior end. Beaks are situated slightly posterior of mid-length, dorsal margins are sloping rather steeply with an equal angle on both sides, and ventral margin is broadly arcuate. The ligament is small and situated in a depressed ligamental groove, postero-umbonal fold is obsolete. Sculpture consists of somewhat widely spaced growth lines crossed by fine incised striae which are concentric on the posterior area, oblique over the remaining part of the shell except on the anterior part of the shell where they become divaricant. The hinge plate is very thin; the right anterior cardinal is trigonal in shape and larger than the right posterior cardinal which is lamellar, bifid, and smaller than the anterior one. In the left valve the anterior cardinal is large and bifid, posterior cardinal is small and lamellar; laterals of both valves are absent. Pallial sinus is large, its dorsal margin is posteriorly high and anteriorly descending until it touches the pallial line immediately behind the anterior adductor muscle scar, ventral margin of it is entirely coalescent with pallial line. Anterior adductor muscle scar is elongate and somewhat smaller than the posterior one which has sub-orbicular shape.

FIGURED SPECIMEN: Australian Museum (Green Island, Queensland); length, 11.0 mm (100); height, 8.4 mm (76.4); thickness, 4.4 mm (40.0).

HABITAT: Known from Queensland, Australia, and New Hebrides.

Subgenus *Loxoglypta* Dall, Bartsch and Rehder

Loxoglypta Dall, Bartsch and Rehder, 1939, p. 192, fig. 153.

TYPE-SPECIES: *Tellina obliquilineata* Conrad (by original designation).

GEOLOGIC RANGE: Recent.

Macoma (*Loxoglypta*) *obliquilineata* (Conrad)
(Pl. 37, figs. 7–11)

Tellina obliquilineata Conrad, 1837, p. 259; Hanley, 1846, p. 254, Pl. 59, fig. 127.

DESCRIPTION: Shell small, thin, white with some rose colored rays, donaciform-elongate, rather inflated, nearly equivalve, strongly inequilateral. Anterior end is long and rounded, posterior end is bluntly truncate, and the beaks are situated a third of the length from posterior end. The antero-dorsal margin is sloping very gently, postero-dorsal margin is steep and slightly excavated behind the beak, ventral margin is nearly straight. The ligament is small and situated in a shallow ligamental groove, postero-umbonal fold is obsolete. Sculpture consists of fine concentric ridges over the entire shell together with fine, oblique, incised rugae superimposed over the concentric sculpture on the middle of the shell extending from the anterior third up to the posterior fold; the concentric ridges tend to be coarse on the posterior area. Hinge plate is very thin, right valve has two well-developed cardinals, the posterior one is bifid; left anterior cardinal is well developed and bifid, left posterior cardinal is thin, lamellar and partly fused with the nymph; laterals of both valves are absent. Pallial sinus is large, its dorsal margin is posteriorly high and anteriorly descending, the anterior end of it is very close to the anterior adductor muscle scar, and ventral margin of it is entirely coalescent with pallial line. Anterior adductor muscle scar is slightly larger and elongate in shape, the posterior one is suborbicular.

FIGURED SPECIMEN: USNM 337341 (Honolulu Harbor, Hawaii); length, 11.4 mm (100); height, 6.5 mm (57.0); thickness, 3.0 mm (26.4).

HABITAT: Hawaiian Islands.

Subgenus *Exotica* Lamy

Exotica Lamy, 1918, p. 117.

TYPE: *Exotica exotica* Lamy (= *triradiata* H. Adams) (by absolute tautonymy).

GEOLOGIC RANGE: Recent.

Macoma (*Exotica*) *triradiata* (H. Adams)
(Pl. 37, fig. 1)

Tellina triradiata H. Adams, 1870, p. 790, Pl. 48, fig. 9.

Exotica exotica 'Jousseaume' Lamy, 1918, p. 117.

DESCRIPTION: A specimen of *M.* (*Exotica*) *triradiata* was not available, and following is the description given by Lamy: "Shell small, solid, donaciform, slightly inflated. Its anterior end is longer and rounded, the dorsal and ventral margins are parallel. Posterior end is very short, obliquely truncate, with acuminate extremity, and bent to the right; this inflection to the right makes the shell appear squeezed at this area. With aid of a high-powered lens, the sculpture on the surface of the shell appears to consist of fine concentric striae which are sharp and closely spaced. The color is an intense red which changes into irregular, small, and broken lines near the umbone; the color is so unstable that it disappears completely when the shell remains for sometime on the sea shore. In the interior, the muscle impressions are well-marked, but pallial sinus is hardly visible. The cardinal area is rather thick and has a shape relative to that of the shell; cardinal teeth are small, and laterals are absent. The ligament is situated in a deep groove, it is small and yellowish."

FIGURED SPECIMEN: Copy of the original figure (Suez, Red Sea); length, 11.5 mm (100); height, 6 mm (52.2); thickness, 4 mm (39.8).

HABITAT: Red Sea.

Subgenus *Salmacoma* Iredale

Salmacoma Iredale, 1929, p. 267.

TYPE-SPECIES: *Salmacoma vappa* Iredale (by original designation).

GEOLOGIC RANGE: Recent.

Macoma (*Salmacoma*) *vappa* (Iredale)
(Pl. 37, figs. 2–6)

Salmacoma vappa Iredale, 1929, p. 267, Pl. 30, figs. 7, 8.

DESCRIPTION: Shell of medium size, rather thin, elongate-ovate, inequivalve, sub-equilateral, white with orange coloring on the umbonal area. Anterior end is rounded and slightly longer, posterior end bluntly truncate; right valve is more inflated than the left valve. Dorsal margins are sloping moderately at an equal angle; ventral margin is convex, twisted, and emarginate at the posterior end. There is a small lunule and a short lanceolate ligament situated in a shallow ligamental groove. Sculpture consists of fine concentric striae which become obsolete on the umbonal area. Hinge plate is narrow, right valve has two cardinals, the anterior one is bifid; left valve has an anterior bifid cardinal and a posterior cardinal which is lamellar and partly fused with the nymph. Laterals of both valves are absent. Pallial sinus is slightly discrepant in two valves, being smaller in the right valve than in the left valve. In the right valve dorsal margin of pallial sinus is high in the middle and anteriorly it is descending until it coalesces with the pallial line 7 millimeters behind the anterior adductor muscle scar; in the left valve the dorsal margin of pallial sinus is ascending on the posterior part, then descending forward until it coalesces with the pallial line 5 millimeters behind the anterior adductor muscle scar. The ventral margin of pallial sinus is entirely coalescent with pallial line. Adductor muscle scars are subequal, the anterior one is elongate and posterior one suborbicular.

FIGURED SPECIMEN: Paratype from Australian Museum (Innisfail, Queensland); length, 31.4 mm (100); height, 23.2 mm (74.0); thickness, 13.3 mm (42.4).

HABITAT: Queensland, Australia.

Subgenus *Ascitellina* Marwick

Ascitellina Marwick, 1928, p. 467.

TYPE-SPECIES: *Ascitellina donaciformis* Marwick (by original designation).

GEOLOGIC RANGE: Lower Miocene.

Macoma (*Ascitellina*) *donaciformis* (Marwick)
(Pl. 37, figs. 12–13)

Ascitellina donaciformis Marwick, 1928, p. 467, figs. 59, 60.

DESCRIPTION: I was unable to obtain a specimen of *M. (Ascitellina) donaciformis;* therefore, I am quoting the original description: "Shell very small, longitudinally oval; beaks behind the middle line, inconspicuous; anterior end long and narrowly rounded, posterior shorter and broader, roundly truncated. Sculpture sharp concentric ridges, about 8 per millimeter on the center of disc, many die out on reaching the posterior area where the remainder become thicker, about 4 or 5 per millimeter. Hinge weak; right valve with a narrowly-triangular, grooved posterior cardinal tooth, and a weak lamellar anterior one; no lateral teeth; ligamental nymph short, sunk below the valve-margin. Pallial line not seen. Valve-margins smooth."

"Height, 6 mm; length, 10 mm; thickness (1 valve), 1.4 mm."

"Locality: Cliffs below Wireless Station, Waitangi." (Bryozoan Limestone). In New Zealand.

FIGURED SPECIMEN: The original figure given by Marwick is reproduced.

Subgenus *Bartrumia* Marwick

Bartrumia Marwick, 1934, p. 10.

TYPE-SPECIES: *Raeta tenuiplicata* Bartrum (by original designation).

GEOLOGIC RANGE: Lower Miocene.

Macoma (Bartrumia) tenuiplicata (Bartrum)
(Pl. 37, figs. 14–15)

Raeta tenuiplicata Bartrum, 1919, p. 97, Pl. 7, figs. 5, 6; Marwick, 1934, p. 10, Pl. 1, figs. 1, 2.

DESCRIPTION: Shell of medium size, thin, inflated, equivalve, inequilateral. Umbones prominent and situated slightly posterior to the middle of the shell, beaks prosogyrate. Anterior end broadly rounded, posterior end subrostrate, bluntly truncate, and sinuous; ventral margin emarginate and broadly arcuate. Sculpture consists of fine concentric lamellae which are obsolete on crossing the posterior area, but become distinct again on the postero-dorsal margin. There are, also, weak, undulating, radiate striae which are better developed only on the posterior area of the shell. Ligamental groove is depressed and extends from the umbone to one-third of postero-dorsal margin. Hinge is narrow and the left valve has two short, weak cardinals, the anterior one is slightly bifid. The laterals of both valves are absent. Pallial sinus scarcely visible, but probably it is broadly rounded, extending to the middle of the shell and coalescing with pallial line below.

FIGURED SPECIMEN: The figures of Marwick (1934, pl. 1) are reproduced here (New Zealand); exterior of left valve × 1; hinge view of left valve × 3.

OCCURRENCE: Mokau beds, New Zealand.

Subgenus *Cymatoica* Dall

Cymatoica Dall, 1889, p. 272.

TYPE-SPECIES: *Tellina undulata* Hanley (by subsequent designation Dall, 1900).

GEOLOGIC RANGE: Recent.

Macoma (Cymatoica) undulata (Hanley)
(Pl. 37, figs. 16–20)

Tellina unaulata Hanley, 1844, p. 72; Hanley, 1846, p. 310, Pl. 59, fig. 107; Reeve, 1867, Pl. 23, figs. 119a, b.

DESCRIPTION: Shell very small, thin, white, rather compressed, elongate-ovate, inequilateral, inequivalve. The anterior end is rounded and slightly longer than the posterior end which is rostrate, sharply truncate, and bent to the right. The antero-dorsal margin is sloping very gently, postero-dorsal margin is steep and slightly con-cave, ventral margin is broadly arcuate and emarginate at the posterior part. The escutcheon extends along more than half of postero-dorsal margin, ligamental groove occupying upper half of it and there is a shallow lunule. The postero-umbonal fold is obsolete. The sculpture is constituted of two elements; one consisting of fine con-centric growth lines, the other is coarse, concentric, slightly curved, and somewhat undulating ribs extending from postero-umbonal fold to the anterior end and ter-minating at the anterior margin. The right posterior and left anterior cardinals well-developed and bifid; right anterior and left posterior cardinals are well-marked, but small, lamellar, and thin. Laterals of both valves are absent. Pallial sinus is small, its dorsal margin nearly straight, the anterior end rounded, and the ventral margin of it entirely free. Adductor muscle scars are small and suborbicular.

FIGURED SPECIMEN: USNM 211454 (La Paz, Baja California); length, 9.5 mm (100); height, 5.0 mm (52.6); thickness, 2.1 mm (22.1).

HABITAT: Gulf of California to Ecuador.

GENUS APOLYMETIS SALISBURY, 1929

DIAGNOSIS: Shell of medium to large size, white, suborbicular, inflated, inequivalve, inequilateral. Anterior end is broadly rounded and long, posterior end is plicate and bluntly truncate. Sculpture is concentric, and a rather short ligament is situated in a depressed ligamental groove. There are two cardinals in each valve and laterals are absent in both valves. Pallial sinus is generally large, and its ventral margin is mostly or entirely coalescent with pallial line.

TYPE: *Tellina meyeri* Dunker (by monotype).

Key to the Subgenera of Genus *Apolymetis*

A. Sculpture of concentric lamellae. . . . . *Apolymetis* s.s.
B. Sculpture of concentric striae.
   a. Anterior end of pallial sinus coalescent with anterior adductor
      muscle scar. . . . . *Leporimetis*
   aa. Anterior end of pallial sinus not coalescent with anterior ad-
      ductor muscle scar.
      b. Anterior end of pallial sinus close to anterior adductor
         muscle scar.
         c. Dorsal margin of pallial sinus high. . . . . *Florimetis*
         cc. Dorsal margin of pallial sinus not high. . . . . *Tellinimactra*
      bb. Anterior end of pallial sinus distant from anterior adductor
        muscle scar. . . . . *Pilsbrymetis*
                                new subgenus

Subgenus *Apolymetis* Salisbury

Capsa Lamarck, 1799, p. 84 (not *Capsa* Bruguiere, 1797, nor *Capsa* Lam., 1818, nor *Capsa* Leach, 1852).

Caspa Bosc, 1802, p. 18; error for *Capsa*.

Metis H. and A. Adams, 1856, p. 399 (not *Metis* Philippi, 1843, nor *Metis* Gistel, 1848, nor *Metis* H. and A. Adams, 1856).

Lutricola 'Blainville' Carpenter, 1863, p. 639 (not Blainville, 1825).

Polymetis Salisbury, 1929, p. 255; new name for *Metis* (not *Polymetis* Walsingham, 1903).

Apolymetis Salisbury, Proc. Malac. Soc. London, vol. 18, p. 258.
TYPE-SPECIES: *Tellina meyeri* Dunker (by montype).
GEOLOGIC RANGE: Recent.

Apolymetis (*Apolymetis*) *meyeri* Dunker
(Pl. 38, figs. 1–3)

Tellina meyeri Dunker in Philippi, 1846, p. 49, Pl. 4, fig. 1; Reeve, 1867, Pl. 30, fig. 167; Römer, 1872, p. 207, Pl. 40, figs. 1–3.

DESCRIPTION: Shell large, white, thin, inflated, suborbicular, inequivalve, inequilateral. Anterior end is longer and broadly rounded, posterior end is bluntly truncate and strongly plicate. Beaks are prosogyrate and located a short distance posterior to the mid-length; ligament is rather short and situated in a depressed ligamental groove. Antero-dorsal margin is convex and excavated immediately in front of the beak, and postero-dorsal margin starts with a smooth convex curve from the apex of the beak; ventral margin is broadly arcuate. There are two cardinals in each valve; right posterior and left anterior cardinals are bifid. Laterals of both valves are absent. Pallial sinus is large, its dorsal margin is very high in the middle, its anterior end is coalescent with anterior adductor muscle scar, and its ventral margin is coalescent with the pallial line. Anterior adductor muscle scar is larger than the posterior one.

FIGURED SPECIMEN: I was unable to obtain a specimen of *T. meyeri*, and the illustrations of this species are copies from Römer's monograph, where the measurements are given as: length, 52.0 mm (100); height, 44.0 mm (84.5); thickness, 17.5 mm (33.7).

HABITAT: Malaysia and Indonesia.

Subgenus *Leporimetis* Iredale

Leporimetis Iredale, 1930, p. 74.
TYPE-SPECIES: *Tellina spectabilis* Hanley (by original designation).
GEOLOGIC RANGE: Recent.

Apolymetis (*Leporimetis*) *spectabilis* (Hanley)
(Pl. 38, figs. 4–8)

Tellina spectabilis Hanley, 1844, p. 141; Hanley, 1846, p. 323, Pl. 65, fig. 254; Reeve, 1866, Pl. 6, fig. 22; Römer, 1872, p. 206, Pl. 39, figs. 10–12.

DESCRIPTION: Shell of medium size, white, rather thin, suborbicular, inflated, inequivalve, strongly inequilateral. Anterior end is long and broadly rounded; posterior end is bluntly truncate, strongly plicate, and slightly bent to right. Umbones are situated one-third of the length from the posterior end; dorsal margins are sloping steeply, the postero-dorsal margin being very steep. Ventral margin is slightly concave and twisted. Postero-umbonal fold is stronger on the right valve than it is on the left; ligament is short, stout, and placed in a depressed ligamental groove. Margin of the left valve, immediately in front of the beak, projects out as a small platform which fits under the margin of right valve when the shell is closed. Sculpture consists of concentric thin lamellae which are separated by interspaces having a width equal to three times the thickness of the lamellae, and these interspaces are lined by submicroscopic striae. The same proportion of lamellae and the interspaces are retained throughout the shell surface, but the sculpture becomes finer as it approaches the beak. There are two rather small cardinal teeth in each valve. Right anterior cardinal is short and pointed; right posterior cardinal is somewhat long, rather thin, and feebly bifid. Left anterior cardinal is thin and slightly bifid; left posterior cardinal is thin, lamellar, and mostly fused with the nymph. Laterals of both valves are absent. Pallial sinus is large, its dorsal margin posteriorly high and anteriorly descending, its anterior end is coalescent with the anterior adductor muscle scar, and its ventral margin is entirely coalescent with pallial line. Anterior adductor muscle scar is elongate and larger than the posterior one which is of suborbicular shape.

FIGURED SPECIMEN: From collections of the Australian Museum (Port Curtis, Queensland); length, 38.2 mm (100); height, 32.1 mm (84.0); thickness, 21.1 mm (55.2).

HABITAT: Philippines to Australia.

Subgenus *Florimetis* Olsson and Harbison

Lutricola Carpenter, 1863, p. 639 (not Blainville, 1825).
Florimetis Olsson and Harbison, 1953, p. 129.
TYPE-SPECIES: *Tellina intastriata* Say.
GEOLOGIC RANGE: Pliocene to Recent.

Apolymetis (*Florimetis*) *intastriata* Say
(Pl. 39, figs. 1–5)

Tellina intastriata Say, 1826, p. 218; Römer, 1872, p. 203, Pl. 1, fig. 10, Pl. 39, figs. 4–6.
Tellina gruneri Philippi, 1845, p. 150.

DESCRIPTION: Shell moderately large, rather thin, broadly ovate, inflated, inequivalve, inequilateral. Anterior end is somewhat longer and broadly rounded, posterior end more sharply rounded and slightly bent to right; set off from rest of shell by angled fold in right valve, and conspicuous furrow in left valve. Umbones are situated a short distance posterior to the mid-length, postero-umbonal fold is prominent, and a short stout ligament is situated in a depressed ligamental groove. Antero-dorsal margin slopes down with a smooth convex curve, postero-dorsal margin is straight; ventral margin is nearly straight. Sculpture consists of fine and occasional obscure radial striae. Cardinals are rather small for the size of the shell; right posterior and left anterior cardinals are bifid and somewhat thin, right anterior cardinal is thin, lamellar, and mostly fused with the nymph. Laterals of both valves are absent. Pallial sinus is large, posterior part of its dorsal margin is high, anterior part is descending, its anterior end is broadly rounded and four millimeters from anterior adductor muscle scar, posterior half of its ventral margin is coalescent with pallial line. Adductor muscle scars are large, the anterior one is elongate and larger than the posterior one which has orbicular shape.

FIGURED SPECIMEN: USNM 438803 (Florida); length, 61.2 mm (100); height, 50.1 mm (81.8); thickness, 27.6 mm (45.1).

HABITAT: Florida, West Indies, to Venezuela.

Subgenus *Tellinimactra* Lamy

Tellinimactra Lamy, 1918, p. 169.
TYPE-SPECIES: *Tellina edentula* Spengler (by monotype).
GEOLOGIC RANGE: Recent.

Apolymetis (*Tellinimactra*) *edentula* (Spengler)
(Pl. 39, figs. 6–8)

Tellina edentula Spengler, 1798, p. 96 (not Broderip and Sowerby, 1829).
Scrobicularia inflata Schumacher, 1817, p. 128.
Lutraria tellinoides Lamarck, 1818, p. 470.
Tellina angulata Römer, 1872, p. 209, Pl. 40, figs. 4–6; Smith, 1891, p. 427.
Tellina lamyi Dautzenberg and Fischer, 1907, p. 224.

DESCRIPTION: Shell large, white, rather thin, inflated, inequilateral, inequivalve. Anterior end is longer and broadly rounded, posterior end is bluntly truncate and slightly twisted to right. Umbones are situated about a third of the length from the posterior end; postero-umbonal fold is stronger on the right valve than on the left. Antero-dorsal margin slopes moderately, postero-dorsal margin steeply; ventral margin broadly arcuate and slightly sinuous at the posterior part. A short and stout ligament is situated in a depressed ligamental groove. Sculpture consists of fine concentric ridges and growth lines. There are two cardinals in each valve, right posterior and left anterior cardinals are bifid; right anterior cardinal is small, the posterior cardinal of left valve is thin and partly fused with nymph. Laterals of both valves are absent. Pallial sinus is large, its dorsal margin is only slightly convex, its anterior end is rounded and about four millimeters from anterior adductor muscle scar, half of its ventral margin is coalescent with pallial line. Anterior adductor scar is elongate, and the posterior one is orbicular in shape.

FIGURED SPECIMEN: Copied from Römer's monograph (Philippines); length, 57.0 mm (100); height, 47.0 mm (82.5); thickness, 20.0 mm (35.0).

HABITAT: Red Sea to Philippines.

Subgenus *Pilsbrymetis*, New Subgenus

DIAGNOSIS: Shell large, white, rather thin, inflated, inequilateral, inequivalve. Anterior end is rounded and longer, posterior end is plicate and bluntly truncate. Dorsal margins are steep; ventral margin is arcuate, twisted in the middle, and sinuous at the posterior part. Postero-umbonal fold is stronger on the right valve than on the left; a short, stout ligament is situated in a depressed ligamental groove. There are two cardinals in each valve, and laterals of both valves are absent. Pallial sinus is small, has a characteristically rectangular shape, and the ventral margin of it is entirely coalescent with pallial line. Posterior adductor muscle scar is larger than the anterior one.

TYPE-SPECIES: *Tellina papyracea* Gmelin.

GEOLOGIC RANGE: Tertiary to Recent.

Apolymetis (*Pilsbrymetis*) *papyracea* (Gmelin)
(Pl. 40, figs. 1–5)

Tellina papyracea Gmelin, 1791, p. 3231.

Tellina lacunosa Lamarck, 1818, p. 530; Hanley, 1846, p. 322, Pl. 45, fig. 252; Römer, 1872, p. 201, Pl. 38, figs. 10–12.

DESCRIPTION: Shell large, white, rather thin, inflated, inequivalve, inequilateral. Anterior end is rounded and longer, posterior end is short and bluntly truncate. Umbones are prominent and situated a short distance posterior to the mid-length; beaks are orthogyrate. Ligament is short, stout, and situated in a depressed ligamental groove. Antero-dorsal margin has a moderate slope, postero-dorsal margin is convex and has a steep slope. The ventro-posterior part of the shell margin is sinuous due to posterior plication. Ventral margin is broadly arcuate and its posterior part is twisted. Right valve has a broad flexure in the middle extending from umbone to ventral margin and this has produced a slight emargination on the margin. Sculpture consists of widely spaced growth lines and fine concentric ridges which are separated by inter-spaces having a width of four times that of the ridges, and the interspaces are lined by very fine concentric striae; the ridges become slightly lamellar and crowded together on the posterior area. Hinge plate is thin; there are two cardinals in each valve, and laterals of both valves are absent. Cardinals of right valve are of about equal size, separated by a deep socket, and the posterior one is slightly bifid. Left anterior lateral is large, bifid, and slightly up-turned; left posterior cardinal is thin, lamellar, and partly fused with the nymph. Pallial sinus is small, its dorsal margin is straight, its anterior margin nearly straight and delineated by a line which is almost vertical, giving the pallial sinus a characteristically rectangular shape. Anterior end of pallial sinus is 15 millimeters from anterior adductor muscle scar, and its ventral margin is entirely coalescent with pallial line. Both adductor muscle scars are elongate, the anterior one is smaller than the posterior one.

FIGURED SPECIMEN: USNM 75880 (Guinea, West Africa); length, 58.0 mm (100); height, 50.0 mm (86.2); thickness, 30.0 mm (51.7).

HABITAT: Guinea, West Africa.

GENUS *MACALIA* H. ADAMS, 1860

Macalia H. Adams, 1860, p. 369.
Tellinungula Römer, 1873, p. 268.

DIAGNOSIS: Shell small to medium size, white, somewhat thick, solid, compressed, suborbicular, inequilateral, subequivalve. Anterior end is long and rounded, posterior end is short and bluntly truncate. Postero-umbonal fold is absent; a short, lanceolate ligament is situated in a depressed ligamental groove, and the beaks are prosogyrate. Sculpture consists of concentric ridges, and fine radial striae. There are two cardinals in each valve, anterior cardinals of both valves are strikingly large; laterals of both valves are absent. Pallial sinus is discrepant in two valves, that of right valve being the smaller. Half of the ventral margin of pallial sinus is coalescent with pallial line. Anterior adductor muscle scar is smaller than the posterior one.

TYPE-SPECIES: *Tellina bruguieri* Hanley (by original designation).

GEOLOGIC RANGE: Recent.

Macalia (Macalia) bruguieri (Hanley)
(Pl. 41, figs. 1–5)

Tellina bruguieri Hanley, 1844, p. 142; Hanley, 1846, p. 321, Pl. 62, fig. 192; Reeve, 1867, Pl. 30, fig. 165; Römer, 1872, p. 268, Pl. 50, figs. 6–9.

DESCRIPTION: Shell of medium size, white, solid, somewhat thick, compressed, trigonal-orbicular, inequilateral, subequivalve. Anterior end is broadly rounded and long, posterior end is short and bluntly truncate. Umbones are situated posterior to the mid-length, beaks are prosogyrate, and postero-umbonal fold is absent. Antero-dorsal margin has a moderate slope; postero-dorsal margin is convex where it is adjacent to the beak but straight for the remaining part and has a steep slope; ventral margin is broadly arcuate. Escutcheon is shallow and poorly defined, and a small ligament is situated in a depressed ligamental groove; lunule is small and narrow. Sculpture consists of fine concentric ridges separated by interspaces having a width of about twice that of the ridges, these are crossed by very fine radial striae which produce a slightly decussate sculpture on the umbonal area. There are two cardinals in each valve, the anterior cardinal of each valve is strikingly large. Right anterior cardinal is large, pointed, and slightly up-curved; right posterior cardinal is small, and bifid. Left anterior cardinal is strikingly large, trigonal, and strongly bifid; left posterior cardinal is thin, lamellar, and mostly fused with the nymph. Laterals of both valves are absent. Pallial sinus is discrepant in two valves, that of right valve being smaller. In the right valve the dorsal margin of pallial sinus is slightly high on the posterior part and steeply descending on the anterior part, its anterior end is acutely rounded and five millimeters from anterior adductor muscle scar; posterior half of its ventral margin is coalescent with pallial line. Pallial sinus of left valve is large, its dorsal margin is high on the posterior part and gently descending on the anterior part, its anterior end is broadly rounded and only two millimeters from anterior adductor muscle scar, posterior half of its ventral margin is coalescent with pallial line. Anterior adductor muscle scar is elongate and smaller than the posterior one which has orbicular shape.

FIGURED SPECIMEN: USNM 17868 (Singapore); length, 33.2 mm (100); height, 28.3 mm (82.8); thickness, 13.7 mm (40.0).

HABITAT: Singapore to Philippines.

GENUS *GASTRANA* SCHUMACHER, 1817

Gastrana Schumacher, 1817, pp. 44, 132.
Diodonta Deshayes, 1846, pl. 68.
Fragilia Deshayes, 1848, p. 552.

DIAGNOSIS: Shell of medium size, white, solid, subtrigonal-ovate, inflated, equivalve, inequilateral. Anterior end is short and broadly rounded, posterior end is narrow and acutely rounded. Dorsal margins are steep, and the ventral margin is broadly arcuate. Umbones are situated anterior to the mid-length, postero-umbonal fold is broad, and a small ligament is situated in a depressed ligamental groove. Sculpture consists of fine concentric ridges and very fine radial striae. There are two cardinals in the right valve, the posterior one is bifid and slightly larger. Left anterior cardinal is large, trigonal, and bifid; the posterior cardinal of the left valve is small, thin, and mostly fused with the nymph. Laterals of both valves are absent. Pallial sinus is of medium size, its anterior end is constricted and acutely rounded, half of its ventral margin is coalescent with pallial line. Anterior adductor muscle scar is smaller than the posterior one.

TYPE-SPECIES: *Tellina fragilis* Linnaeus (by subsequent designation Stoliczka, 1870).

GEOLOGIC RANGE: Pliocene to Recent.

Gastrana (Gastrana) fragilis (Linnaeus)
(Pl. 41, figs. 6–10)

Tellina fragilis Linnaeus, 1758, p. 674; Römer, 1872, p. 276, Pl. 52, figs. 4–7.
Venus fragilis Fabricius, 1780, p. 413.
Psammotaea tarentina Lamarck, 1818, p. 518.
Petricola ochroleuca Lamarck, 1818, p. 503.
Psammobia fragilis Turton, 1822, p. 88, Pl. 7, figs. 11–12.
Fragilia fragilis Deshayes, 1848, p. 561, Pl. 68.
Gastrana fragilis Mac'Andrew, 1857, p. 105.
Capsa fragilis Weinkauff, 1867, p. 60.

DESCRIPTION: Shell of medium size, white, solid, inflated, subtrigonal-ovate, equivalve, inequilateral. Anterior end is short and broadly rounded; posterior end is narrow and acutely rounded. Umbones are situated slightly more than a third of the length from anterior end; dorsal margins are steep, and ventral margin is broadly arcuate with a slight emargination at the posterior end. Postero-umbonal fold is broad, and a small ligament is situated in a depressed ligamental groove. Sculpture consists of fine concentric ridges separated by interspaces having a width of about twice that of the ridges on the central area. These ridges and interspaces are crossed by very fine radial striae which are better visible on the central area of the shell. Right valve has two sharply pointed, somewhat small cardinals separated from each other by a deep socket; the posterior one is bifid and slightly larger than the anterior one. On the left valve anterior cardinal is large, trigonal, strongly bifid, and set out by a deep socket on each side. Left posterior cardinal is small, thin, and mostly fused with the nymph. Pallial sinus is of medium size, its dorsal margin descending, its anterior end acutely rounded and extending forward only slightly beyond the mid-length, and posterior half of its ventral margin is coalescent with pallial line. Anterior adductor muscle scar is elongate and smaller than the posterior one which is larger and has orbicular shape.

FIGURED SPECIMEN: USNM 178525 (Algiers); length, 32.0 mm (100); height, 23.9 mm (74.6); thickness, 12.8 mm (40.0).

HABITAT: Mediterranean Sea, and in the Atlantic from Greenland and Scandinavia to Senegal, West Africa.

Subgenus *Heteromacoma* Habe

Heteromacoma Habe, 1952, p. 218.
Sinomacoma Yamamoto and Habe, 1959, p. 102.
TYPE: *Tellina irus* Hanley (by original designation).
GEOLOGIC RANGE: Pliocene to Recent.

Gastrana (Heteromacoma) irus (Hanley)

Tellina irus Hanley, 1844, p. 166; Hanley, 1846, p. 319, Pl. 60, fig. 145.
Fragilia yantaiensis Crosse and Debeaux, 1863, p. 78.
Heteromacoma irus Habe, 1952, p. 218, figs. 542, 543; Habe and Ito, 1965, p. 149, Pl. 51, figs. 13, 14.
Sinomacoma yantaiensis Yamamoto and Habe, 1959, p. 102, Pl. 9, figs. 4, 5.

COMMENT: After I had finished the manuscript, Number 3 of the *Genera of Japanese Shells* by T. Habe came out in which he has proposed a new genus *Heteromacoma* with *T. irus* Hanley as the type. Because I have no opportunity to make any studies on *T. irus*, and its synonyms and other aspects of the problem, I am simply giving the original description of *T. irus* Hanley:

DESCRIPTION: "Ovate or obovate, solid, coarse, subventricose, subequilateral, dull dirty white both within and without, the surface roughened by minute concentrically arranged but unconnected elevated wrinkles (reminding one of those upon *Rugosa* but on a much smaller scale), which however are usually (except in portions) abraded; front dorsal edge convex after passing the small but distinct lunule; ligamental edge not greatly convex, and decidedly sloping so as to form an angle with the more or less arched ventral margin; beaks inclining to the rounded extremity of the shorter anterior side; ligament sunken but distinct; a large bifid and a simple scarcely rudimentary tooth in one valve, two rather strong diverging obtuse ones in the other."

HABITAT: Is given as "Guinea" which is obviously an error; the actual habitat extends from Alaska to Japan and California.

REMARKS: T. Habe states that: "This group of species (*T. irus* Hanley, *T. japonica* Martens, and *T. oyamai* Kira MS.) were put into *Gastrana* Schumacher, 1817, but the characteristics of teeth are altogether different." The differences in the character of dentition between *T. irus* Hanley and *T. fragilis* Linnaeus is in the size of left posterior cardinal; in the former it is small, whereas in the latter it is very small. I do not believe anatomically there is such a difference between the two as to be of generic magnitude. As I consider the genus a category with broad latitude, I place *Heteromacoma* Habe as a subgenus under *Gastrana* Schumacher.

COMMENT: The genus *Sinomacoma* was proposed by Yamamoto and Habe in 1959 on the assumption that *Macoma irus* Hanley is the North American species synonymous with *Macoma inquinata* Deshayes. As far as I know, *Gastrana (Heteromacoma) irus* (Hanley), a solid shell with fine concentric rugulose sculpture, is not found in the eastern Pacific. The statement by Salisbury (1934, p. 85) that *irus* Hanley is synonymous with *inquinata* Deshayes needs confirmation; the specimens labeled *inquinata* that I have examined appear to belong to *Macoma* s. str.

It should be noted that the same figures are used in the original proposal of both *Heteromacoma* and *Sinomacoma* Habe and Ito (1965, p. 149) recognize the identity of both of these two genera.

Appendix

HARALD A. REHDER

In order to bring this paper as much up to date as possible by including all genera and subgenera proposed since Dr. Afshar completed his manuscript and thus make it more useful to persons consulting this work, I am listing all super-specific genera proposed since 1952 that were not covered in Dr. Afshar's manuscript. They are arranged for convenience in alphabetical order.

In each case, the bibliographic reference and type-species designation is followed by the diagnosis copied verbatim from the original publication. In some cases, this is followed by comments of my own.

SUPPLEMENT

Subgenus *Hertellina* Olsson

Hertellina Olsson, 1961, Panamic-Pacific Pelecypoda, p. 409.
TYPE-SPECIES: *Tellina (Scissula) nicoyana* Hertlein and Strong, 1949.
DIAGNOSIS: Shell subelliptical, thin, nearly equilateral (*Sanguinolaria*-like in shape), the anterior side a little longer than the posterior. Surface smooth with indistinct growth lines and a heavier sculpture of concentric ribbons between incised lines which are a little oblique to the growth lines and margin of the valve. Hinge plate narrow and delicate, the teeth small; right valve has two, small, grooved cardinal teeth and a long, narrow, anterior lateral placed fairly close to the cardinals and a distant posterior lateral tooth; left valve has two cardinal teeth of which the anterior one is bifid, the other simple. Pallial line is large and deep, not reaching to the anterior adductor scar and separated from it by considerable space, the lower limb confluent with the pallial line below.

Externally shaped like *Sanguinolaria*, but with hinge characters of *Eurytellina*.

Named to honor Dr. Leo George Hertlein of the California Academy of Sciences in recognition of his many major contributions to the paleontology and malacology of the Pacific region.

Tellina (Hertellina) nicoyana (Hertlein and Strong)
(Pl. 44, figs. 7–10)

Tellina (Scissula) nicoyana Hertlein and Strong, 1949, Zoologica, vol. 34, pt. 2, pp. 85, 86, pl. 1, figs. 23–26.

97

DESCRIPTION: The shell is relatively small, elongately ovate to elliptical, thin, mildly convex, and of a pale rose or pink color. The surface is smooth, the lines of growth weak but with a stronger sculpture of concentric ribbons between evenly spaced incised lines which begin on the posterior-middle side of the disk and run diagonally to the margin in the middle zone and concentric or parallel to the margin on the anterior side.

Length 34.4 mm, height 19 mm, diameter 7.8 mm. Ballena Bay, Gulf of Nicoya, Costa Rica. Holotype, Calif. Acad. Sci.

This is evidently a rare species known only from a few specimens. The shape is like that of a small *Sanguinolaria* but the hinge is that of a Tellina. A specimen labelled *Sanguinolaria panamensis* Dall (evidently an unpublished name) is in the collection at the U. S. National Museum from Bay of Panama (USNM 96361). I have a single valve which was dredged off Zorritos, Peru.

HABITAT: Costa Rica to northern Peru. Panama: Panama Bay (USNM). Peru: Zorritos.

GEOLOGIC RANGE: Recent.

Subgenus *Elpidollina* Olsson

Elpidollina Olsson, 1961. Panamic-Pacific Pelecypoda, p. 407

TYPE-SPECIES: *Tellina decumbens* (Carpenter, 1865)

DIAGNOSIS: Shell with thin, subtrigonal valves, thin and rather inflated, subequal, the posterior side shorter and pointed, and hardly flexed. The hinge provided with both cardinal and lateral teeth, the cardinals usually small and of which the left anterior and the right posterior teeth are bifid, the others much smaller and simpler. The lateral teeth are fairly large in the right valve, much smaller in the left; the anterior lateral tooth is placed near but not actually in contact with the cardinal tooth, the posterior lateral tooth more distant and beyond the end of the ligament scar. The ligament is external, its scar long and narrow. The pallial sinus is large and deep, highest under the beak and extending across to connect with the anterior adductor scar; its lower limb is fully confluent with the pallial line. Surface smooth except for minute lines of growth.

Tellina (Elpidollina) decumbens (Carpenter)
(Pl. 44, figs. 1–4)

Angulus amplectans Carpenter, 1863, Rept. British Assoc. Adv. Sci., p. 669, nude name.

(*Tellina*) *Angulus decumbens* Carpenter, 1865, Proc. Zool. Soc. London, pp. 278, 279. Reprinted 1872, Smith. Misc. Coll., No. 252, pp. 271, 272, Panama.

Tellina peasii Sowerby, 1868, Conch. Icon., vol. 17, Tellina, pl. 49, fig. 288.

Tellina (Moerella) decumbens (Carpenter), Myra Keen, 1958, Sea Shells of Tropical West America, p. 170, fig. 395.

DESCRIPTION: Shell of medium size (to about 47 mm.), obliquely subovate to subtrigonal, the anterior side longer, higher, its dorsal margin somewhat expanded, the posterior side shorter, pointed, pinched but with hardly any flexing. The valves are nearly equal, moderately convex, especially along the anterior umbonal slope, slightly depressed on the posterior-ventral slope. Valves are relatively thin, smooth except for fine, growth concentrics, the color for the most part white, but some shells have the umbones and the shell cavity within flushed with pink; periostracum is thin, light-cream color.

Length 40.8 mm., height 38.2 mm., diameter 15.5 mm.

Young shells may resemble *T. lineata* Turton of the Caribbean but are more elliptical, thinner, and more convex. Most specimens are white, but occasional specimens have the umbones stained with rose or pink. This appears to be a localized species; it is common at Old Panama.

HABITAT: Panama: Old Panama.

GEOLOGIC RANGE: Recent.

Subgenus *Acorylus* Olsson And Harbison

Acorylus Olsson and Harbison, 1953, Acad. Nat. Sci. Phila. Mon. No. 8, p. 128.

TYPE-SPECIES: *Tellina (Moerella) suberis* Dall, 1900.

DIAGNOSIS: Shell small, solid, moderately convex, obliquely subovate, with a posterior flexure, best developed in the right valve. Hinge stout; in the right valve with two cardinal teeth and a central socket, and two strong lateral teeth equidistant from the cardinals, each forming the lower border of a deep, lateral socket into which the margin of the opposite valve fits. Left valve has a single cardinal tooth flanked by a socket on each side, no true lateral teeth. The external ligament is attached to a long scar lying above that of the resilium and extending from the tip of the beak to the end of the posterior lateral socket. Pallial sinus large, deep, reaching to the anterior adductor scar and widely confluent with the pallial line below.

Dall and most later authors have referred the species of this group to *Moerella* Fischer, the type of which is *Tellina donacina* Linne, a species not related to the forms here under consideration. Gardner has shown that the name '*Moerella*' should be used for the American species formerly placed in *Angulus*.

Quadrans (Acorylus) suberis Dall
(Pl. 43, figs. 1–4)

Tellina (Moerella) suberis Dall, 1900b, Trans. Wagner Free Inst. Sci., vol. 3, pt. 5, p. 1031, Pl. 46, fig. 25.

Tellina (Acorylus) suberis Olsson and Harbison, 1953, Acad. Nat. Sci. Phila., Mon. No. 8, p. 128, Pl. 14, figs. 5, 5a, 5b.

DESCRIPTION: Shell small, obliquely subovate, the anterior side large, convex and rounded at end, the posterior side shorter, descending, slightly flexed. Surface smooth, polished or finely concentrically striated, the lines strongest near the ventral margin and on the sides.

This species resembles the *T. gouldii* Hanley of the Recent fauna of Florida, but differs by its smaller size, less regular form, and its posterior side is less strongly flexed. Very common at St. Petersburg.

OCCURRENCE: Caloosahatchie formation, Florida.

GEOLOGIC RANGE: Pliocene.

GENUS BATHYTELLINA HABE

Bathytellina Habe, 1958, Publ. Seto Marine Biol. Lab., Vol. 7, No. 1, p. 46.

TYPE-SPECIES: *Bathytellina citrocarnea* Kuroda and Habe, 1958; plate 43, fig. 6.

DESCRIPTION: Shell elongate oval, thin but rather solid, moderately inflated, orange rose, especially deep in color at the flexuous postero-dorsal area; beak small, situated at the posterior third of dorsal margin; the antero-dorsal margin being longer than the postero-dorsal; the anterior margin rounded and the posterior margin slightly rostrated and weakly truncated at the extremity; the hinge plate of the right valve with two cardinal teeth, the anterior tooth is stronger than the posterior and bifurcate and anterior and posterior lateral teeth are far from the cardinal teeth; in the left valve no lateral teeth present; mantle line with a deep posterior sinuation.

Length 19.4 mm, height 12.1 mm and breadth 6.2 mm (type specimen).
Length 18.2 mm, height 11.5 mm and breadth 5.5 mm (paratype specimen).
TYPE LOCALITY: Tosa Bay, Shikoku (collected by Mr. A. Teramachi).
LOCALITIES: St. 216 (off Kannoura, Lochi Pref., 274 m); St. 220 (Tosa Bay, 234 m); Sts. 437 and 439 (off Goto Islands, 307 m and 155 m) and St. 475 (off Tsushima, 192 m).
DISTRIBUTION: Kyushu; Shikoku and Honshu (north to Sagami Bay).
GEOLOGIC RANGE: Recent.
REMARKS: This new deep sea tellinid species has a different hinge armature from the genera *Fabulina* and *Moerella* despite the close resemblance of shape. Therefore this is proposed as a new generic name *Bathytellina*.

Subgenus *Lyratellina* Olsson

Lyratellina Olsson, 1961. Panamic-Pacific Pelecypoda, p. 383.
TYPE-SPECIES: *Tellina lyra* Hanley.
DIAGNOSIS: Valves elliptical in shape, white or glassy, equal, depressed, or slightly convex; the beaks median, prosogyrate, the posterior side rounded, unflexed, but with a narrow, strongly sculptured rostral area. The surface is sculptured with strong, concentric ridges between wide, flat interspaces. The ligament is external but lies partly immersed in the hinge plate, the margin of the shell rising above the ligament scar. The hinge is normal and strong; in the right valve, there is a single, bifid, cardinal tooth, its posterior arm large, inclined, bordered on the sides by sockets; there is a lateral socket on each side, the anterior one placed a little closer to the cardinal teeth, the lower rim of each socket is enlarged to form a tooth; the left valve has a single, small, narrow, bifid, cardinal tooth bordered on the sides by sockets, the lateral teeth are merely slight enlargements of the hinge margin. There is a small, sunken lunule, larger in the left valve and not set apart by a line. The escutcheon is narrow and deep, the margin of the valve arising above it as a narrow wing.
REMARKS: The type species (*Tellina lyra*) was referred by Dall and some later authors to *Macaliopsis* Cossmann, 1886 (type species, *Tellina barrandei* Deshayes, an Eocene shell from the Paris Basin) but the resemblance of the American species to the Eocene shell is merely superficial. Two fossil forms of *Lyratellina* are now known, *L. protolyra* (Anderson) from Miocene of Colombia, and *L. aequizonata* (Pilsbry and Olsson) from the Pliocene of Ecuador.

There are two species in the Panamic-Pacific region.
I. Beaks high and sharp, the valve margin in front deeply excavated. Concentric lamellae of the surface sculpture of medium size, spaced about a millimeter apart. *L. lyra*
II. The beaks are lower, the margin in front not deeply excavated. Concentric lamellae much finer and spaced three to a millimeter.   *L. lyrica*

Arcopagia (*Lyratellina*) *lyra* (Hanley)
(Pl. 45, figs. 1–4)

Tellina lyra Hanley, 1844, Proc. Zool. Soc. London, p. 68, Tumbez.-Hanley, 1846, Thes. Conch., vol. 1, Tellina, p. 271, pl. 62, fig. 187. Sowerby, Conch. Icon., vol. 17, Tellina, pl. 36, fig. 203.
Tellina (*Macaliopsis*) *lyra* Hanley, Maxwell Smith, 1944, Panamic Marine Shells, p. 64, fig. 842.-Hertlein and Strong, 1949, Zoologica, vol. 34, pt. 2, no. 9, p. 81.
DESCRIPTION: Shell elliptical, compressed, thin, the anterior side higher, more rounded, the posterior side with its dorsal margin descending. Rostral area as long as the posterior side, narrow, excavated and well sculptured. Sculpture consists of small,

regular, thin, sharp concentrics separated by flattened interspaces about a millimeter wide. The pallial sinus is large, its highest point forming an acute angle a short way in front of and on level with the middle of the posterior adductor scar, its end rounded, its lower limb confluent with the pallial line.

A rare species. The largest specimen seen, a right valve from the beach at Fort Amador, Balboa, measures: length 56.2 mm, height 39.6 mm, diameter 6 mm

HABITAT: Lower California to northern Peru. Off San Salvador and Guatemala (H. and S.). Bucaro. Panama Fort Amador beach, Panama Canal Zone: Sua; Mompiche; Ecuador: Tumbez Peru: (type locality).

Subgenus *Nelltia* Stephenson, 1952

Nelltia Stephenson, 1952. U. S. Geol. Survey Prof. Paper 242, p. 113

TYPE-SPECIES: *Nelltia stenzeli* Stephenson (original designation)

DIAGNOSIS: "Shell broadly subelliptical in outline, compressed. Beaks small, non-prominent, prosogyrate, situated in advance of the midlength. Radiating sculpture apparently wanting. Ligament opisthodetic, external, extending nearly half way to the rear; groove deep, nymph thick and prominent. Hinge of left valve with two cardinal teeth, the anterior one prominent, trigonal, bifid, forming a thick-limbed inverted V, the posterior one simple, of medium thickness, oblique, partly fused against the anterior end of the nymph; the anterior cardinal is bounded on either side by a narrow, deep socket, the two sockets connected above the crest of the inverted V. A thin, prominent lateral tooth extends from near the top of the anterior cardinal, forward for 5 or 6 mm in adults, and is separated from the sharp outer margin of the shell by a deep, narrow channel; posterior lateral short, weak, situated just below the posterior end of the ligamental groove. Hinge of right valve, as seen on the holotype only, with two cardinal teeth. The anterior one is rather poorly preserved, is elevated, oblique toward the front and short, but may be partly broken away; posterior cardinal oblique to the rear; of medium length and thickness, apparently simple; the separating socket is trigonal and deep. A deep anterior lateral socket separates a pair of claspers, the inner element of which is strong and prominent and the outer weak and fused against the inner margin of the shell; the posterior lateral dentition consists of a short, strong tooth separated from the lower end of the ligamental groove by a channel of medium width, is rounded on the inner terminus, lies in a nearly horizontal position, and extends a little short of the midlength."

"Compared with the Recent *Tellina radiata* Linne, the genotype of Tellina Linne, the type-species of this genus is much less elongated, has the beak in front of midlength, lacks a posterior plication and external radial lining, is not flexed at the rear, and is markedly different in its dentition. The anterior cardinal in the left valve of *T. radiata*, though bifid, is less isolated and less sharply V-shaped; the posterior cardinal is much weaker; and the anterior lateral dentition is distant, short, and weak; the beak is back of the midlength. Another striking difference is the exaggerated attenuation of the pallial sinus in *T. radiata*, the terminus of which almost reaches the anterior adductor scar."

Linearia (*Nelltia*) *stenzeli* Stephenson
(Pl. 42, figs. 1–5)

Nelltia stenzeli Stephenson, 1952, U. S. Geol. Survey Prof. Paper 242, p. 113, pl. 28, figs. 9–13

DESCRIPTION: "Shell of medium size, depressed convex, subelliptical in outline, slightly inequilateral, nearly equivalve; the species exhibits marked individual variation in outline in shells from the same piece of rock; some shells are proportionately higher with respect to length and some are much lower than the holotype. Beaks small, non-prominent, incurved, prosogyrate, situated about two-fifth the length of the shell from the anterior extremity. Anterodorsal margin gently descending, gently arches; anterior margin rounded less than a semicircle; ventral margin very broadly rounded, rising steeply at the front and rear; posterodorsal margin very broadly arched. Surface with fine, more or less irregular growth ridges and gentle undulations, with an increase in coarseness toward the margins and especially along the posterodorsal margin."

"Dimensions of the holotype, a right valve: Length 38.2 mm, height 27.3 mm, convexity about 6 mm. An internal mold proportionately higher and shorter than the holotype measures, length 36 mm, height 28 mm; an internal mold proportionately lower and more elongated than the holotype measures, length 33 mm, height 20 mm."

"Ligament external, 8 or 9 mm long in the holotype; groove narrow, deeply impressed; nymph rather thick and prominent. Hinge of right valve with two diverging cardinal teeth, not seen perfectly preserved, separated by a deep trigonal socket. The anterior cardinal appears to be short, rather prominent, and oblique forward, but may be partly broken away; the posterior cardinal is not complete but is rather thin, about twice as long as the anterior cardinal, is oblique backward, and is bordered behind by a narrow, shallow socket. Immediately in front of the anterior cardinal is a well-developed lateral tooth about 5 mm long in the holotype, paralleled above by a deep, narrow socket, which in turn is paralleled by a very weak lateral tooth fused against the inner margin of the shell, the two laterals forming a pair of claspers. Just below the posterior end of the ligamental groove is a short well-developed lateral, which is also bordered above by a narrow channel. The hinge of the left valve, as seen in two shells, neither of which is complete, presents one large, short, trigonal, bifid, inverted V-shaped anterior cardinal tooth and an oblique, narrow, posterior cardinal, the two separated by a narrow, deep socket; in front of the anterior cardinal is a deep, narrow socket that connects above the tooth with the other cardinal socket. A well-developed anterior lateral is present opposite the claspers of the right valve; the posterior lateral is quite weakly developed as seen on one of the specimens. Inner margin smooth. Pallial sinus of moderate depth, punctate, rounded on the inner end, with axis nearly horizontal."

FIGURED SPECIMEN: Figures copied from original reference.

HABITAT: Tarrant and Denton Counties, Texas.

GEOLOGIC RANGE: Late Cretaceous (Woodbine Formation).

Subgenus *Simplistrigilla* Olsson

Simplistrigilla Olsson, 1961. Panamic-Pacific Pelecypoda, p. 390.

TYPE-SPECIES: *Strigilla strata* Olsson, 1961.

DIAGNOSIS: Shell small, subovate, convex, the surface marked with strongly oblique incised lines or sulci crossing the shell from one margin to the other but without any line of flexure or a change in direction.

Strigilla (Simplistrigilla) strata
(Pl. 43, fig. 5)

DESCRIPTION: The shell is small, rounded, subovate, white, relatively solid in texture and moderately convex. The umbones are submedian, the beaks at the end bluntly trigonal. The nepionic surface of the umbone is relatively large and sculptured with hairlike concentrics; this type of sculpture ends sharply and on the succeeding adult portion, the sculpture is produced by rather wide, inclined bands set apart between sharply incised lines which run obliquely across the face of the disk from the posterior or ventral margins to the anterior-dorsal side, the result a rather coarse pattern; these bands are continuous and show little or no flexing or any change of direction along the posterior and anterior submargins.

Length 7.2 mm, height 6.8 mm, semidiameter 3.1 mm (a left valve). Punta Blanca, Ecuador. Holotype, ANSP 218950.

A rare and unusual species easily known by its sculpture.

OCCURRENCE: Panama southward to Ecuador. Panama: El Lagartillo. Ecuador: Punta Blanca.

HABITAT OF TYPE-SPECIES: Panama to Ecuador.

Subgenus *Iraqitellina* Dance And Eames

Iraqitellina, Dance and Eames, 1966, p. 37.

TYPE-SPECIES: *Iraqitellina iraqensis* Dance and Eames.

Moderately small, umbones pointed and situated behind midline at about one-third of shell length; posterior region subrostrate but not carinate, with a broad, well-marked radial depression just anterior to the rostration; anterior half of shell with strong and closely-spaced concentric threads, alternate threads fading out near the mid-line leaving only half as many, more widely-spaced threads on the posterior half; valve margins smooth internally; nymph short; cardinal tooth 3a simple and thin, 3b grooved; lateral tooth P.I. small and distinct, no anterior lateral tooth; pallial sinus long, reaching a point nearly below the anterior adductor muscle scar, its upper arm descendent anteriorly.

REMARKS: This genus seems to be intermediate between the genus *Tellinarius* and the genera *Angulus* and *Tellinides*. Compared with *Tellinarius* it is more inflated, has a reduced number of concentric threads on the posterior half of the surface, is subrostrate but not carinate, has a deeper and more descendant pallial sinus, and lacks a lateral tooth A.I. Compared with *Angulus* and *Tellinides*, each of which has a similar pallial sinus, it is strongly ornamented, more inflated, subrostrate, more asymmetrical and has a lateral tooth P.I. but no A.I.

Barytellina (*Iraqitellina*) *iraqensis* Dance and Eames

Iraqitellina iraqensis Dance and Eames, 1966, p. 37, Pl. 3, figs. 2, 3.

DESCRIPTION: (Holotype, a right valve). Moderately small and nearly colorless. Transversely ovate-subtriangular, subrostrate but not carinate posteriorly, with a broad well-marked radial depression anterior to the rostration. Anterior portion well-inflated, ventral plane of commissure flexuous. Prodissoconch minute, smooth. Umbo small, pointed, moderately prominent, situated behind mid-line at about two-fifths the length of the shell. Antero-dorsal margin moderately long and descendent, practically straight, slightly concave medially. Anterior end well-rounded. Ventral margin gently convex, a little ascendent posteriorly, the plane of commissure bulging posteriorly. Posterior end subrostrate, situated moderately low down, gently convex. Postero-dorsal margin a little shorter than antero-dorsal, more steeply descendent, gently concave near umbo, gently convex distally. Ornamented with distinct concentric threads; alternate threads fade out at about the mid-line, posterior half of surface having only half as many, more widely-spaced threads. Valve margins smooth internally. Nymph short. Cardinal tooth 3a thin and oblique, 3b grooved; a short, distinct posterior lateral tooth P.I. situated about two-fifths of the distance along the postero-dorsal margin from the umbo; no anterior lateral tooth. Pallial sinus long, deep, rather narrow, extending to a position beneath the anterior adductor muscle scar, its upper arm a little concave and descendent anteriorly.

DIMENSIONS: Holotype (BMNH 196524). Length, 4.5 mm; height, 3.1 mm; thickness (one valve), 0.8 mm.

MATERIAL: Gurmat Ali, left bank, 16 ft. below surface of Alluvium (type locality), holotype and two paratypes, other paratypes from Gurmat Ali, left bank, at 20 and 56 ft., and from Gurmat Ali, right bank, at 56 ft. below surface of Alluvium.

Subgenus *Austromacoma* Olsson

Austromacoma Olsson, 1961, Panamic-Pacific Pelecypoda, p. 419.
TYPE-SPECIES: *Macoma constricta* (Bruguiere, 1792).

Macoma (*Austromacoma*) *constricta* (Bruguiere)
(Pl. 45, figs. 5–8)

DIAGNOSIS: Shell subovate, fairly convex, generally thin, the posterior side weakly or strongly flexed toward the right, the left valve as a result somewhat larger and more inflated than the right. Two cardinal teeth in each valve, the posterior right and the anterior left teeth bifid; lateral teeth absent. Ligament entirely external, attached to a long, slender scar behind the cardinal teeth; no nymphal ridge. The anterior adductor scar is long and narrow, lucinoid, with the pallial line attached to its lower end and follows close to the ventral margin. Pallial sinus large, somewhat discrepant in the two valves, widely coalescent with the pallial line below. Surface white, marked with fine, hairlike concentric threads and covered with a thin light-colored periostracum.

Typical *Austromacoma* has a large pallial sinus, high and pointed under the beak, and connected with the anterior adductor scar at its lower end. The genus appears to be restricted to the Caribbean.

GEOLOGIC RANGE: Recent.

Subgenus *Ardeamya* Olsson

Ardeamya Olsson, 1961, Panamic-Pacific Pelecypoda, pp. 417–418.
TYPE-SPECIES: *Tellina columbiensis* Hanley, 1844.
DIAGNOSIS: Shell elongate-elliptical, subequivalve, inequilateral, compressed or slightly convex, thin, smooth, white. Posterior side somewhat shorter, pointed or wedge-shaped at the end, straight or with the right valve weakly flexed. Hinge plate small, forming a narrow roof or shelf above the umbonal cavity and bearing two small, weak cardinal teeth in each valve, the left posterior and the right anterior cardinal teeth bifid, the others small and simple; no laterals but the valve margins may be grooved and overlapping. Ligament external, posterior of the beak, the resilifer small, obliquely wedge-shaped and showing a chalky surface texture. Pallial sinus ample, angled, highest in front of the posterior adductor, joining and becoming confluent with the pallial line near the anterior one-third. Adductor scars subequal, shiny. Surface smooth, often polished, the growth lines small and close, covered by a thin, drab-colored periostracum.

Macoma (*Ardeamya*) *columbiensis* (Hanley)
(Pl. 44, figs. 5–6)

Tellina columbiensis Hanley, 1844, Proc. Zool. Soc. London, p. 71. Hanley in Sowerby, 1846, Thes. Conch., vol. 1, p. 307, no. 166 (as 165), pl. 65, fig. 246. Monte Christi.
DESCRIPTION: Shell elongate-elliptical, compressed or slightly convex, thin or fragile, white, smooth, with a length up to about 80 mm, but usually smaller. Anterior side is the longer, high, the small beaks somewhat pointed and projecting, the ventral margin widely rounded, the posterior side shorter, pointed or wedge-shaped at the end, straight or in the right valve sometimes slightly flexed. Surface smooth and marked with minute lines of growth, the whole covered with patches of a thin, cream or drab-gray periostracum.

Length 60 mm, height 34.9 mm, diameter of a right valve 5 mm. Tumbez, Peru.
Length 68.2 mm, height 39.7 mm, diameter of a right valve 5 mm. Punta Picos, Peru.

This is a lovely species, distinguished by its depressed, subelliptical shape and white, shiny valves. Although the beaks are small, they rise sharply above the general outline of the valves.

HABITAT: Panama to northern Peru. Panama: Lagartillo near Las Tablas; San Carlos. Ecuador: Manta; Mompiche. Peru: Tumbez; Zorritos; Punta Picos.

GEOLOGIC RANGE: Recent.

Subgenus *Bendemacoma* Eames

Bendemacoma Eames, 1957, Bull. British Museum (Nat. Hist.), Geology vol. 3, no. 2' p. 66.

TYPE-SPECIES: *Peronaea nigeriensis* Newton (by original designation).

DIAGNOSIS: Of large-medium size, rather thick-shelled, transversely oval-sub-triangular, length considerably exceeding height, inflation moderate. Beaks small, moderately prominent, prosogyrous. Surface ornamented with accentuated growth-lines, posteriorly with two very vague carinae. Escutcheons long, narrow. Lunule narrow, shorter than escutcheon, limited by a fine incised line. Left valve: 2a vertical, distinctly grooved dorsally; 2b moderately oblique, very thin and lamellar, simple; no lateral teeth. Right valve: 3a rather solid, directed moderately forward grooved dorsally; 3b a little longer than 3a, directed backward moderately obliquely, well grooved; no lateral teeth. Nymph long and rather narrowly tongue-shaped, its upper part rising slightly for a short distance from the posterior adductor impression, then gently descendent for most of its length, narrowly rounded at its apex, its lower part coalescent with the pallial line for nearly half its length. Valve margins smooth.

Macoma (*Bendemacoma*) *nigeriensis* (Newton)
(Pl. 43, figs. 7–9)

Peronaea nigeriensis Newton, 1922, Geol. Survey Nigeria Bull., vol. 3, p. 91, Pl. 11, figs. 1–3.

Macoma (*Bendemacoma*) *nigeriensis*, Eames, 1957, Bull. British Museum (Nat. Hist.), Geology, vol. 3, no. 2, p. 66.

DESCRIPTION: Shell trigoniform, broad and ovately oblong, nearly equilateral, more or less smooth, plano-convex, obtusely angulate posteriorly, anterior border the most pronounced and pursuing an obliquely linear direction to the ovatiform marginal curvature, ventral margin elongately curved; teeth consisting of two divergent cardinals each provided with a groove at the summit, the anterior tooth thickening with age and becoming coalescent with the strong cardinal fulcrum, lateral teeth obsolete; the adductor scars are unequal, the anterior being elongate while the other is much rounder, both are concentrically and radially striated; an obscure and, more or less swollen rib proceeds obliquely along each side of the floor of the valve from beneath the cardinal arch to the inner margins of both the adductor scars; pallial line with an extensive elliptical indentation extending nearly horizontally beyond the middle of the valve and more or less parallel with the ventral margin, sometimes the upper line of the sinus or identation gradually and obliquely ascends to the margin of the posterior adductor scar; sculpture exhibits fine concentric lineations grouped into regular periods of growth, crossed by innumerable delicate radial striations, whilst the posterior angulation is furnished with two or three nearly obsolete radial ribs; close vertical striations cover the wide, internal marginal surface of the ventral region.

DIMENSIONS:

	Adult
Height	52 millimeters
Length	70 millimeters
Diameter (united valves).	30 millimeters

REMARKS: This large tellinoid shell is well represented in the collection; the valves being generally in an isolated condition although some of them are capable of being accurately paired. The specimens agree with the contour lines of *Tellina pellucida, T. zitteli, T. grandis,* and *T. latissima,* from the Egyptian Eocenes, as delineated by figures and descriptions published by Mayer-Eymar, and continued by Oppenheim, the latter regarding such species as one form of shell under the specific name of *zitteli.* In other important respects, however, the Nigerian valves differ from the Egyptian shell. They are of much deeper construction measuring more than double the diameter of the Egyptian valves, an adult form of which, as quoted by Oppenheim, yielding a height and length respectively of 56 and 75 millimeters shows a diameter of only 13 millimeters for their united valves. There are, besides, no lateral teeth which, according to Mayer-Eymar, characterize the Egyptian shell. So far as sculpture is concerned, both shells exhibit a more or less zonal arrangement in the structure of the concentric striations; the closely-set radial striations are, however, absent in Egyptian valves, although some distant obscure radail ribs are noticeable on the posterior area of a valve as represented by Oppenheim's largest figure of *Tellina zitteli.* The Nigerian valves are thick and strong, being mostly mineralized and consisting of a siliceo-calcareous structure. Relics of a periostracum are frequently preserved. This form is regarded as belonging to Poli's genus *Peronaea* on account of its non-rostrate character, its more or less trigonal shape, and its possession of a nearly horizontal pallial sinus, this latter character being much rounder and more expansive in *Arcopagia,* with which Oppenheim associated Mayer-Eymar's Egyptian shell of *Tellina zitteli.*

The characters of the hinge and pallial sinus indicate that this species is not a *Peronaea;* it appears to belong to a new subgenus of *Macoma,* characterized by the shape of the shell, the grooved 3a, and the form of the pallial sinus" (Eames, 1957, p. 66–67).

OCCURRENCE: Near Ameki, Nigeria.

GEOLOGIC RANGE: Upper Eocene (Bartonian).

Subgenus Psammothalia Olsson

Psammothalia Olsson, 1961, Panamic-Pacific Pelecypoda, p. 416–417.

TYPE-SPECIES: *Tellina cognata* C. B. Adams, 1852.

DIAGNOSIS: The shell is subrhomboidal or *Psammobia*-like in shape, with flattened umbones and the beaks placed a little behind the middle, the two ends of the valves almost alike, except that the shorter posterior side is more depressed and its margin more widely truncated. The hinge is provided with small cardinal teeth, but there are no laterals; the left valve has a single, narrowly bifid, cardinal tooth with a socket on each side: the right valve has two, small, bifid, cardinal teeth. The ligament is external, attached to a long, narrow, grooved scar along the upper side of a nymphal ridge. The adductor scars are large and nearly equal, placed high in the interior of the valve. The pallial sinus is deep, its upper limb almost reaching across to the anterior adductor scar, rounded at its end and joined with the pallial line below for about half its length. The external surface is usually smooth and polished, marked lightly with concentric lines of growth and diagonally across these a set of incised lines (*Scissula*-like) which cover most of the disk in front of a smooth band along the posterior-umbonal slope.

The generic relations of *Tellina cognata*, the type species, has long been controversial; it was referred to *Psammobia* by Reeve; to *Quadrans* by Dall; and to *Scissula* by Hertlein and Strong. The squat rhombic shape of its valves and oblique surface sculpture is strongly suggestive of the Psammobiidae but the nymph is tellinoid, small, and narrow. The cardinal teeth are all bifid and there are no laterals.

Macoma (*Psammothalia*) *cognata* (C. B. Adams)
(Pl. 42, figs. 6–10)

Tellina cognata C. B. Adams, 1852, Ann. Lyceum Nat. Hist. New York, vol. 5, pp. 503, 545, no. 459.-Turner, 1956, Occas. Papers on Mollusks, Mus. Comp. Zool., vol. 2, no. 20, p. 38, pl. 18, figs. 9, 10.

Tellina concinna C. B. Adams, 1852, op. cit., pp. 504, 546, no. 461. Turner, 1956, op. cit., p. 41, pl. 18, figs. 16, 17.

Psammobia casta Reeve, 1857, Conch. Icon., vol. 10, *Psammobia*, pl. 8, fig. 55. Not *Tellina casta* Hanley, 1844 (Guatemala).

Macoma (*Psammacoma*) *cognata* (C. B. Adams), Maxwell Smith, 1944, Panamic Marine Shells, p. 65, fig. 849.

Tellina (*Scissula*) *cognata* C. B. Adams, Hertlein and Strong, 1949, Zoologica, vol. 34, pt. 2, no. 9, p. 85.

DESCRIPTION: Shell with oblong, rectangular valves, slightly convex to depressed; the anterior end obliquely rounded; the posterior side shorter, depressed, its margin straight as if crudely: runcated. Surface color is principally white except for the umbones which may be shaded lightly with salmon-pink or brown; the interior often pink. The surface is generally smooth and polished; the growth lines showing indistinctly and sometimes under a lens; fine radial striations may often be seen; but the principal sculpture is formed by a series of evenly spaced, diagonal, incised lines which begin in front of a smooth ray bounding the posterior-umbonal angle. The periostracum is straw-gray in color, thin and preserved only on live shells.

Length 53.7 mm, height 35.8 mm, diameter 15.1 mm. Tumbez, Peru.

Length 65.5 mm, height 41 mm, semidiameter 9 mm, a right valve, Zorritos, Peru.

This species, long misunderstood and poorly represented in most museum collections, is now known to be common and widely distributed. A smooth form or one without the obliquely incised lines was named *concinna* by C. B. Adams. It is a common fossil in the Pliocene of Ecuador. The same species or a similar one has recently been discovered in the Gulf of Mexico and seen by the author (Steger collection, Tampa, Florida).

HABITAT: Mexico to northern Peru. For northern records, see Hertlein and Strong. Panama: Bucaro; San Carlos. Panama Canal Zone: Venado Beach. Ecuador: Sua, Charapota, Playas. Peru: Tumbez; Zorritos; Boca Pan.

Plates

The illustrations are of natural size unless otherwise indicated.

PLATE 1.

1. Exterior of left valve.
2. Exterior of right valve.
3. Dorsal view of the valves.
4. Interior of right valve.
5. Interior of left valve.

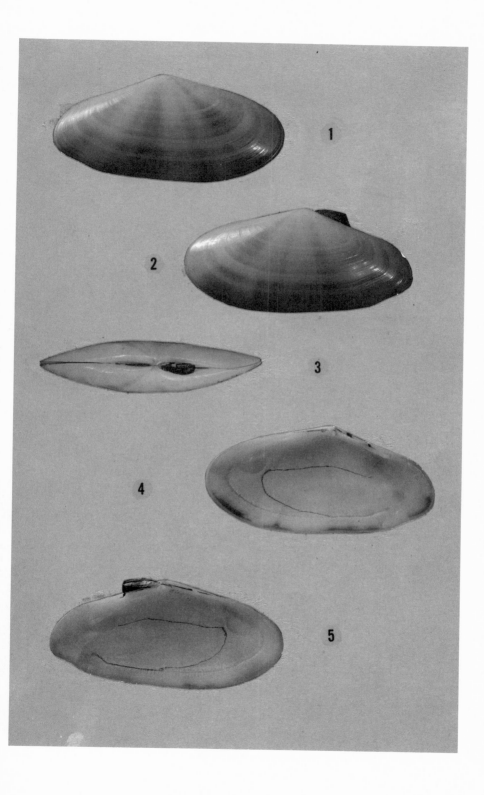

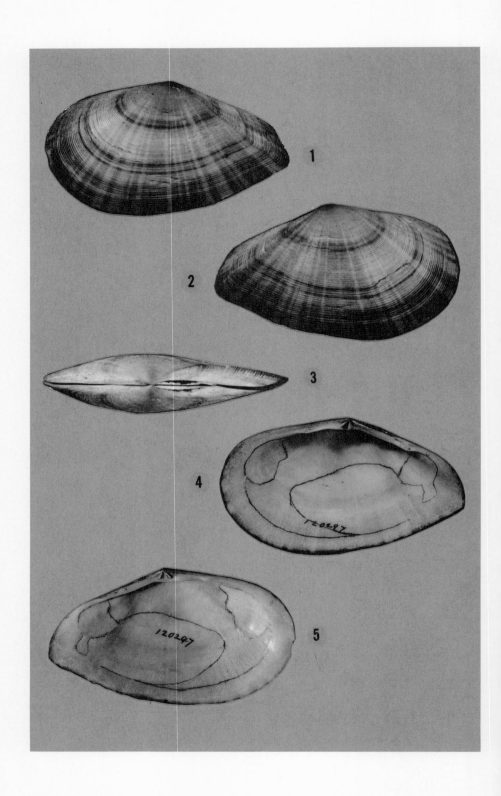

PLATE 2.

1. Exterior of left valve.
2. Exterior of right valve.
3. Dorsal view of the valves.
4. Interior of right valve.
5. Interior of left valve.

PLATE 3.

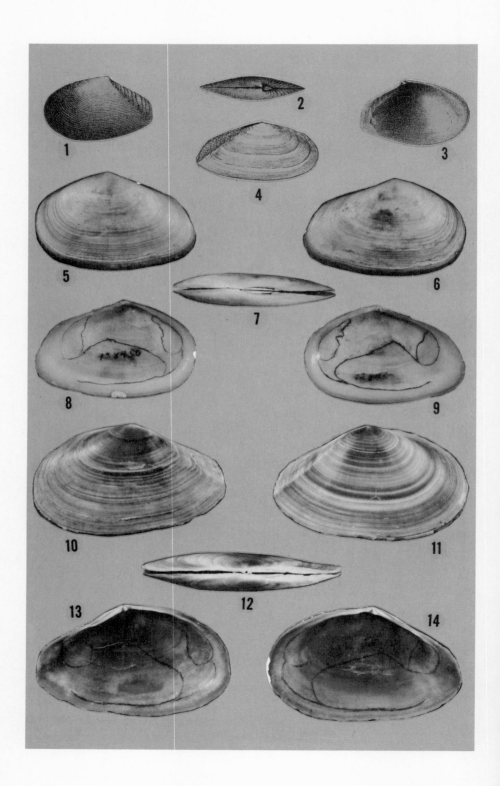

PLATE 4.

PLATE 5.

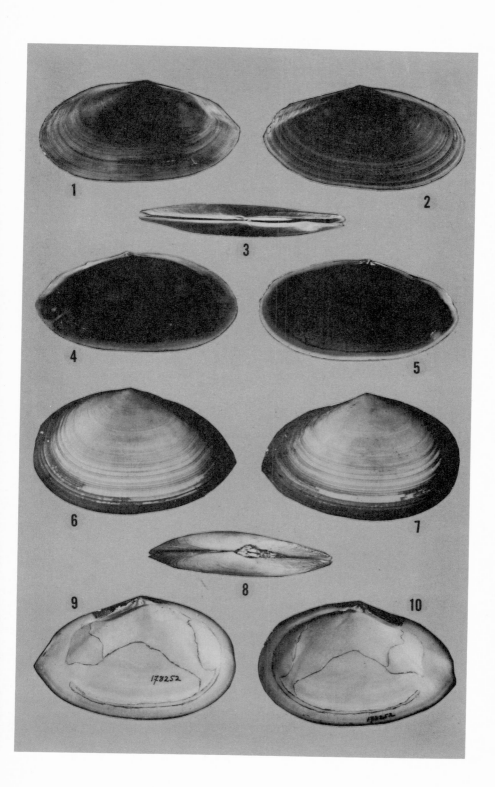

PLATE 6.

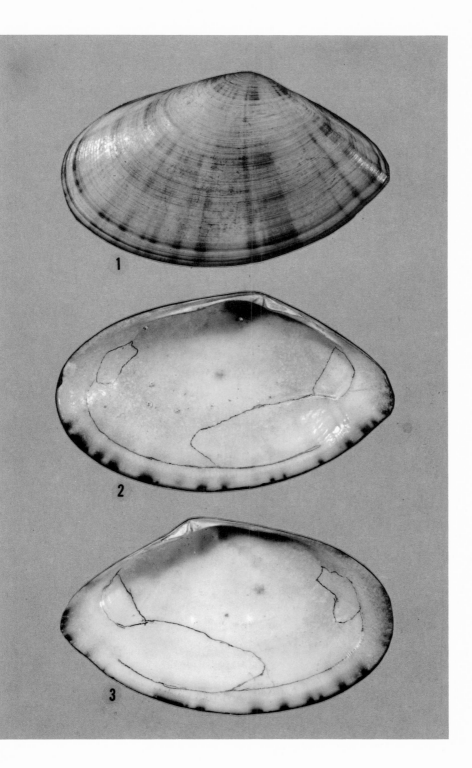

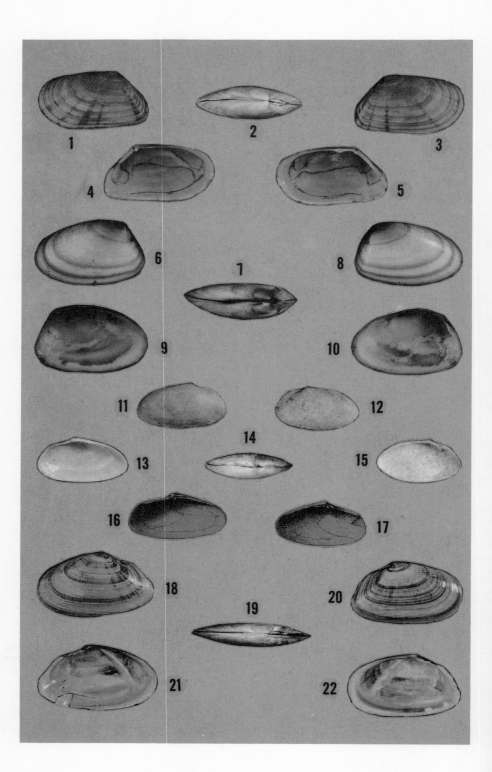

PLATE 8.

PLATE 9.

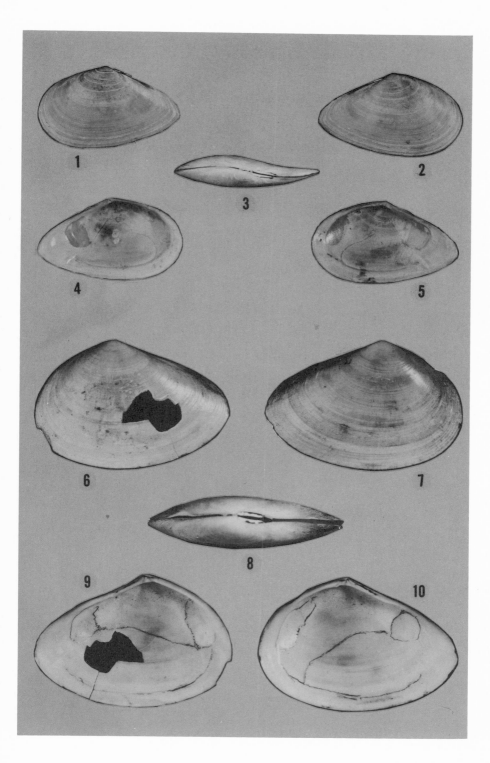

PLATE 11.

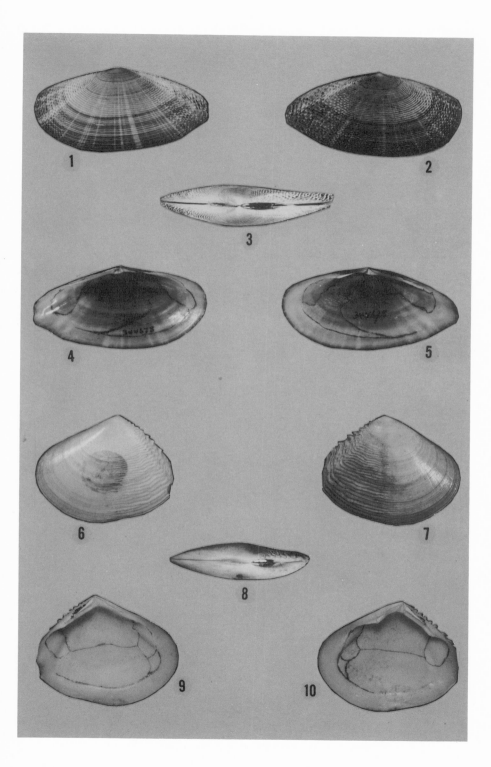

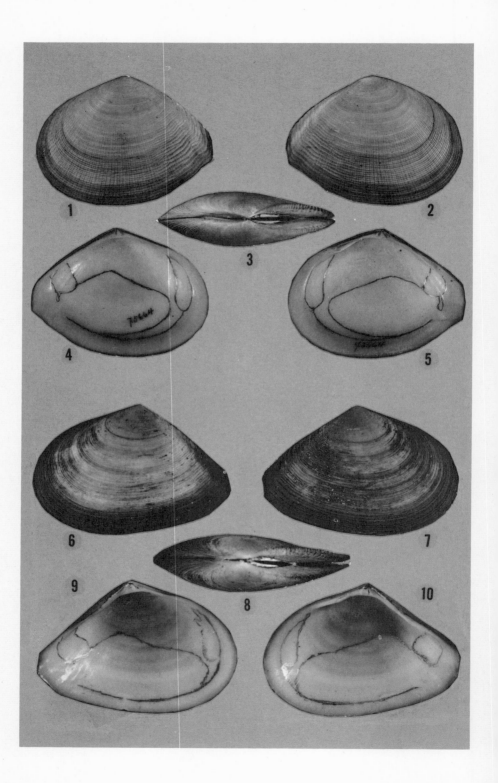

PLATE 12.

PLATE 13.

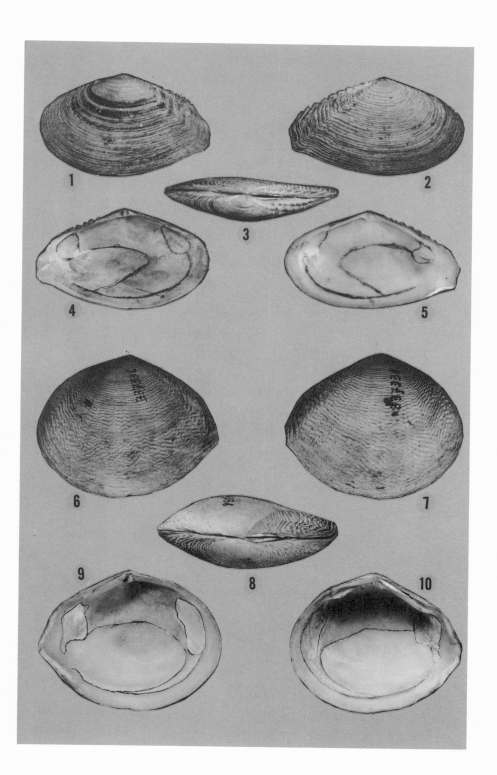

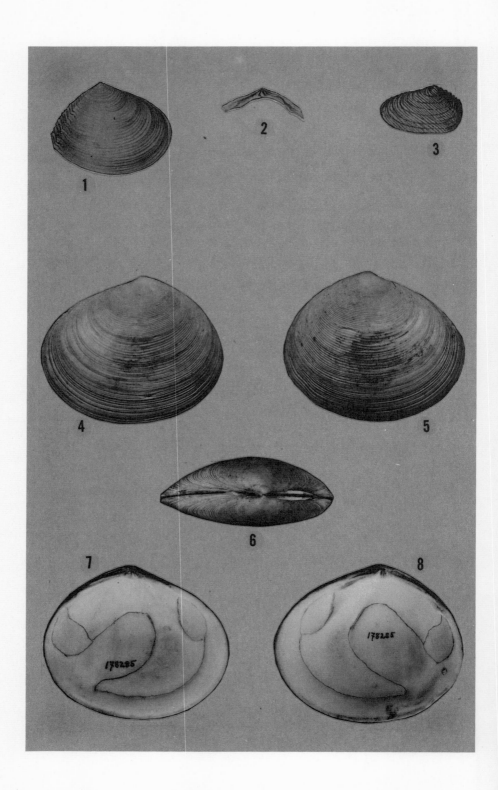

PLATE 14.

PLATE 15.

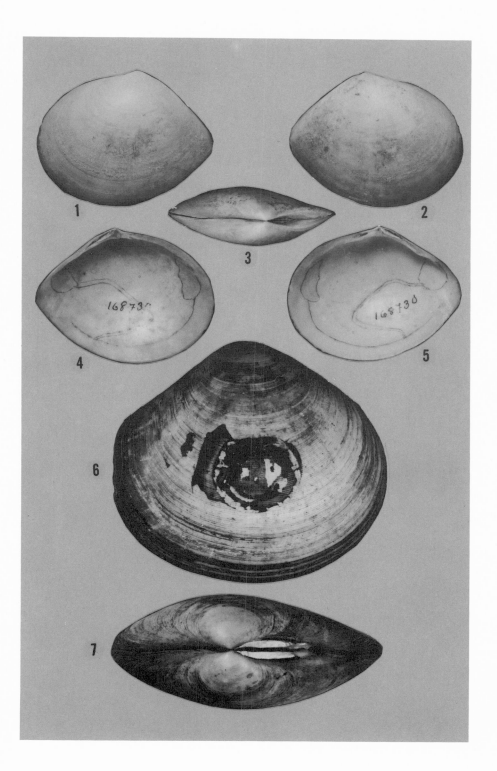

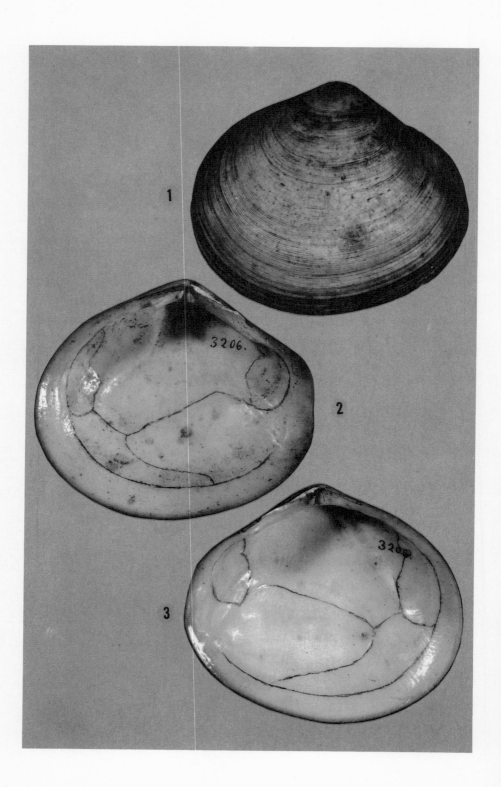

PLATE 16.

PLATE 17.

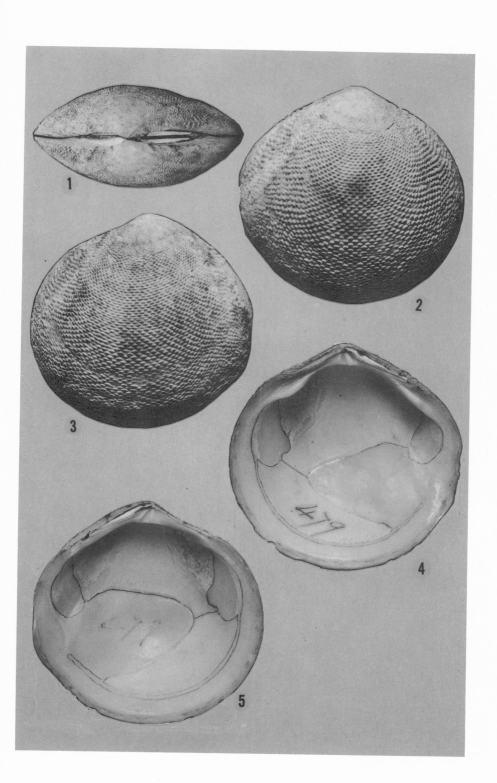

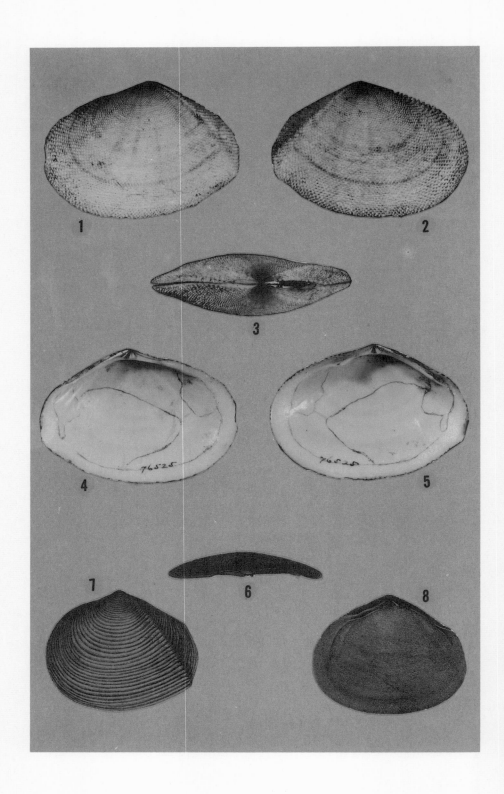

PLATE 18.

PLATE 19.

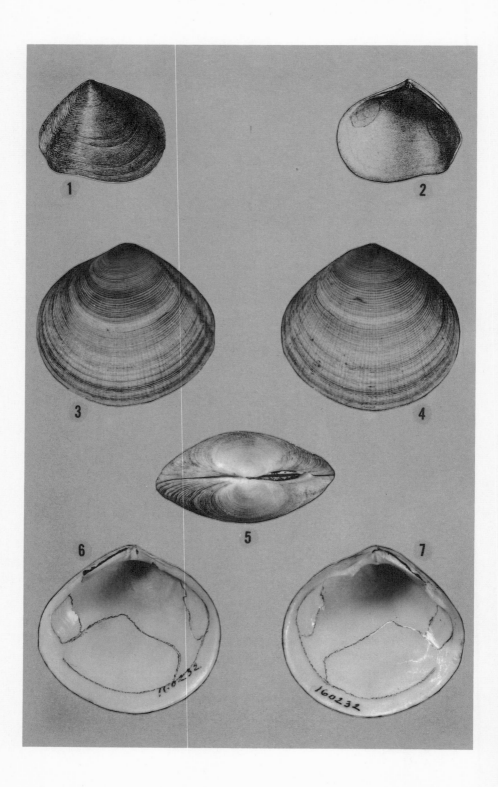

PLATE 20.

Figures

Page

Arcopagia (*Sinuosipagia*) *colpodes* (Bayan). 50
1. Exterior of right valve.
2. Interior of right valve.

Arcopagia (*Pseudarcopagia*) *victoriae* (Gatliff and Gabriel). 50
3. Exterior of left valve.
4. Exterior of right valve.
5. Dorsal view of the valves.
6. Interior of left valve.
7. Interior of right valve.

PLATE 21.

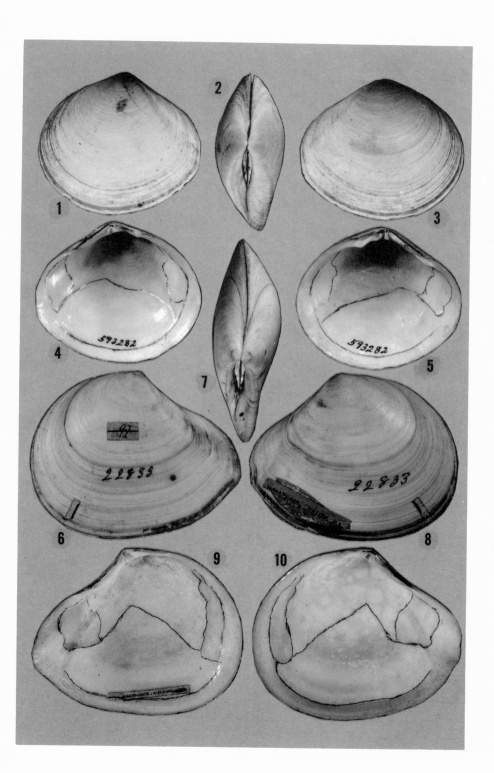

PLATE 22.

PLATE 23.

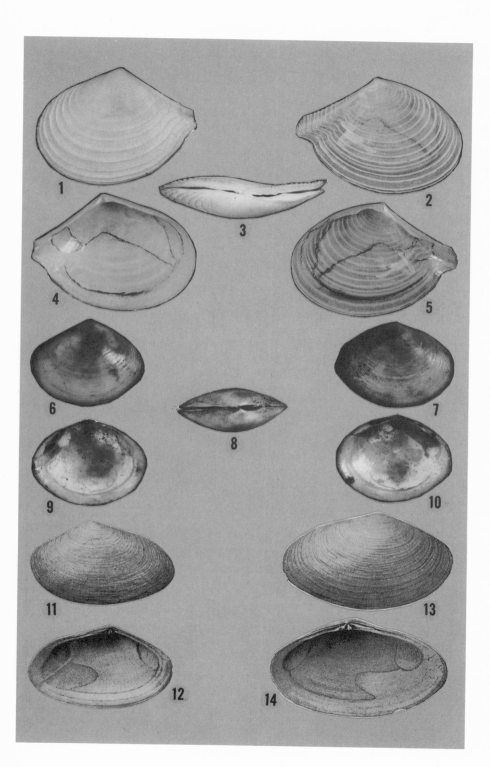

PLATE 24.

PLATE 25.

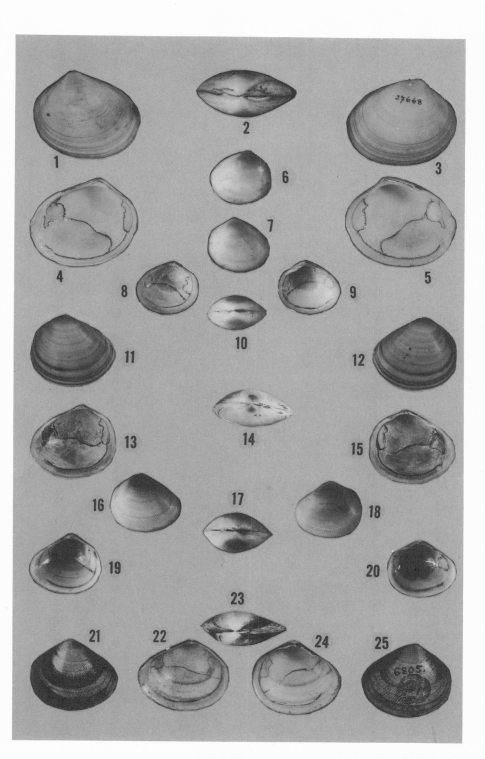

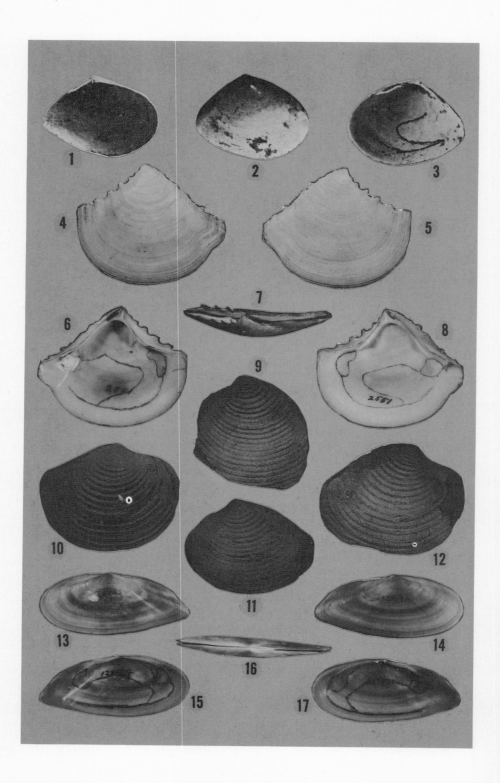

PLATE 26.

PLATE 27.

1. Exterior of left valve.
2. Exterior of right valve.
3. Dorsal view of the valves.
4. Interior of right valve.
5. Interior of left valve.

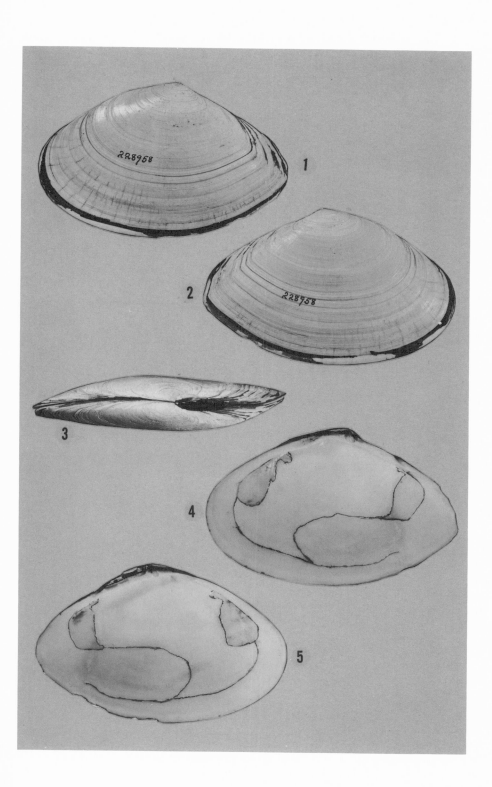

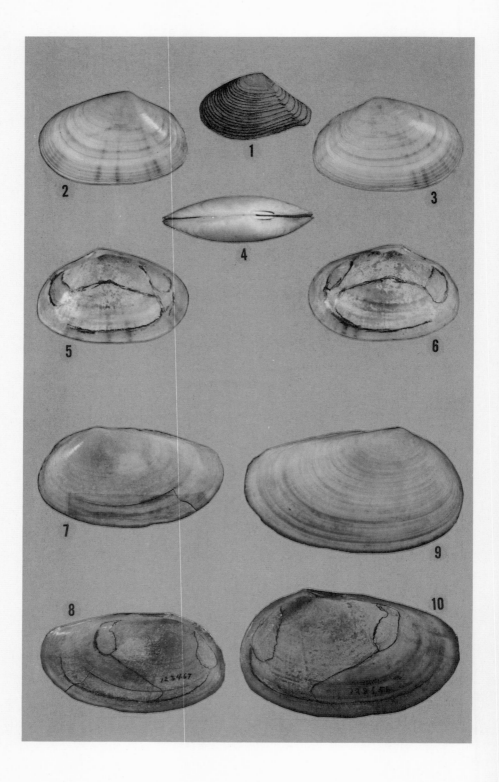

PLATE 28.

PLATE 29.

1. Interior of right valve.
2. Interior of left valve.
3. Exterior of right valve.

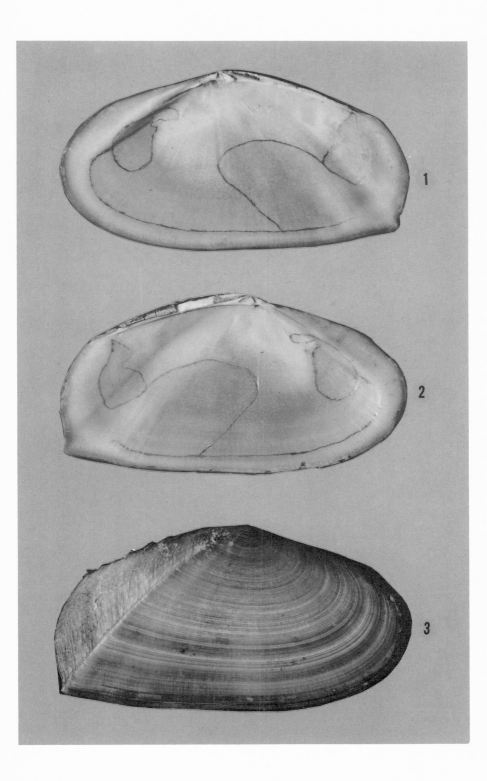

1

2

3

PLATE 30.

PLATE 31.

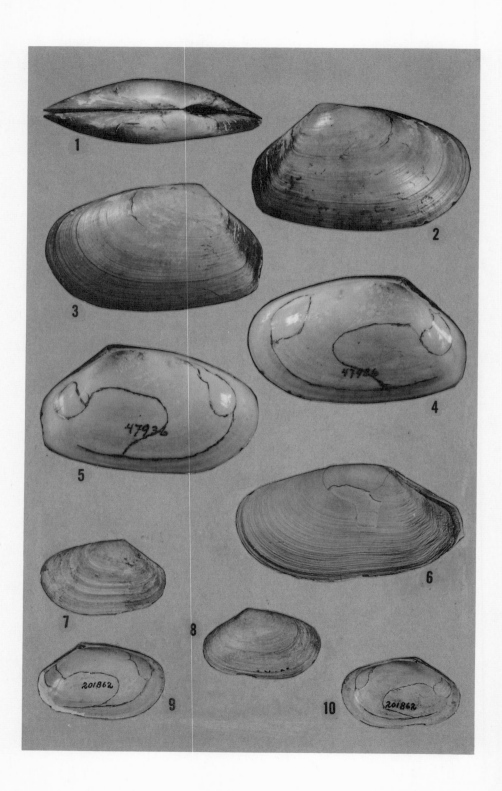

PLATE 32.

PLATE 33.

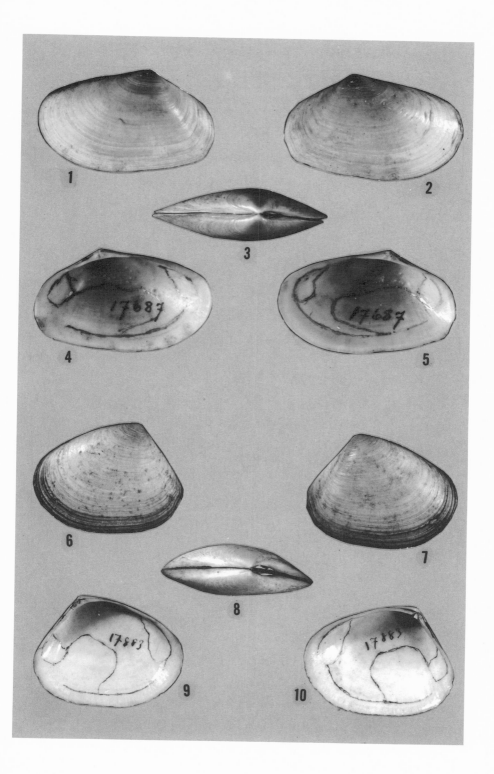

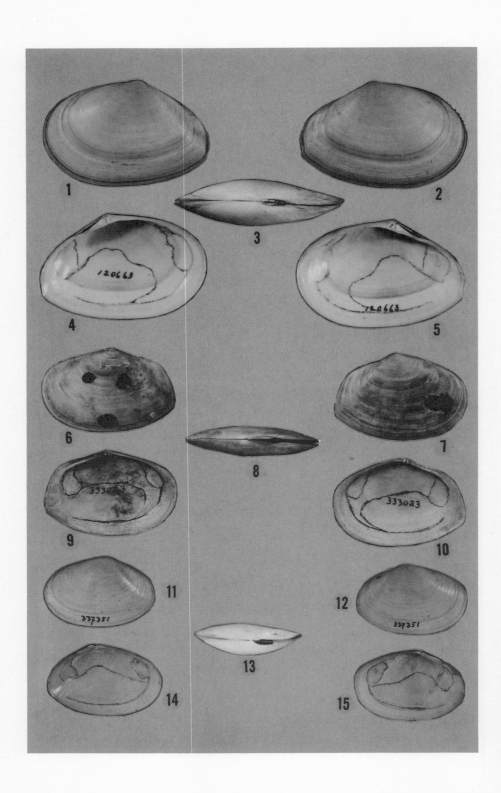

PLATE 34.

PLATE 35.

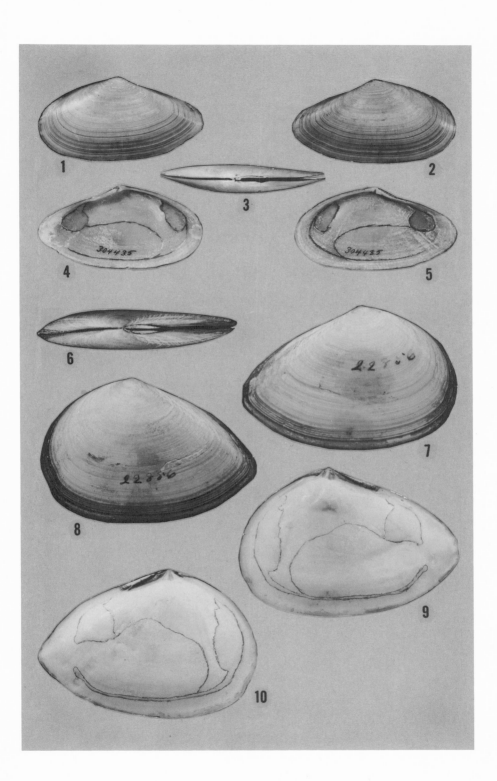

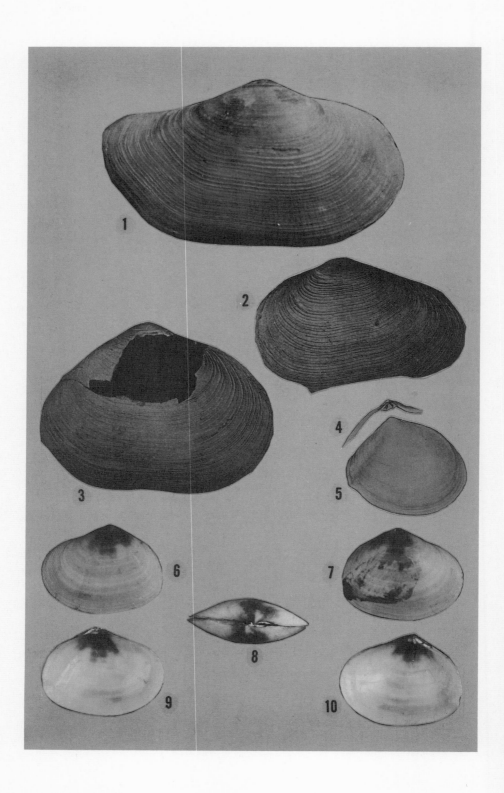

PLATE 36.

PLATE 37.

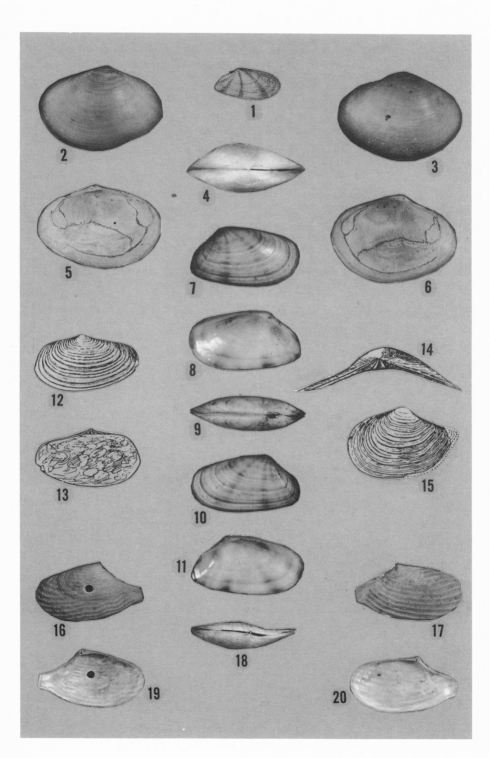

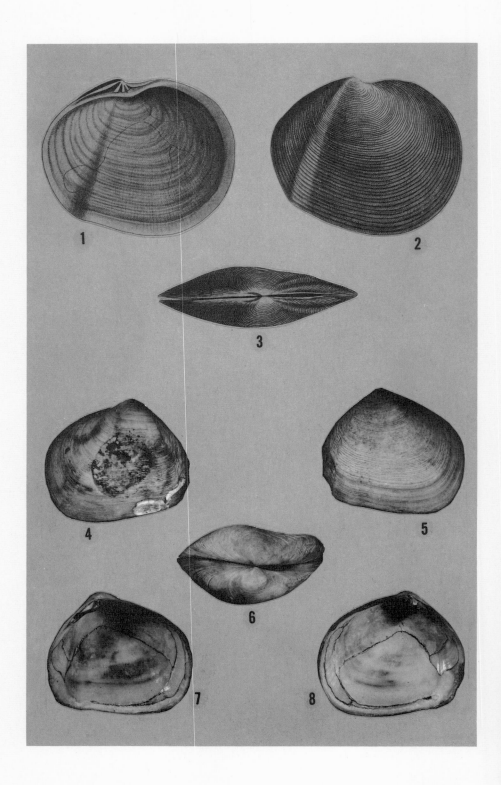

PLATE 38.

1. Interior of left valve.
2. Exterior of right valve.
3. Dorsal view of the valves.

4. Exterior of left valve.
5. Exterior of right valve.
6. Dorsal view of the valves.
7. Interior of left valve.
8. Interior of right valve.

PLATE 39.

PLATE 40.

PLATE 41.

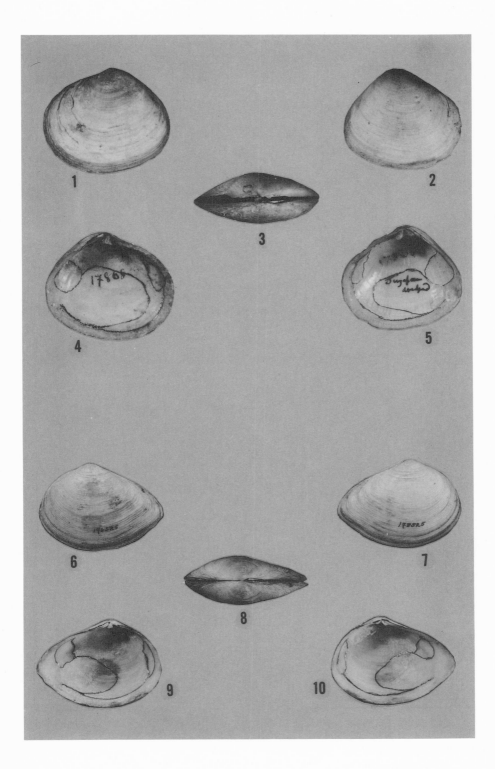

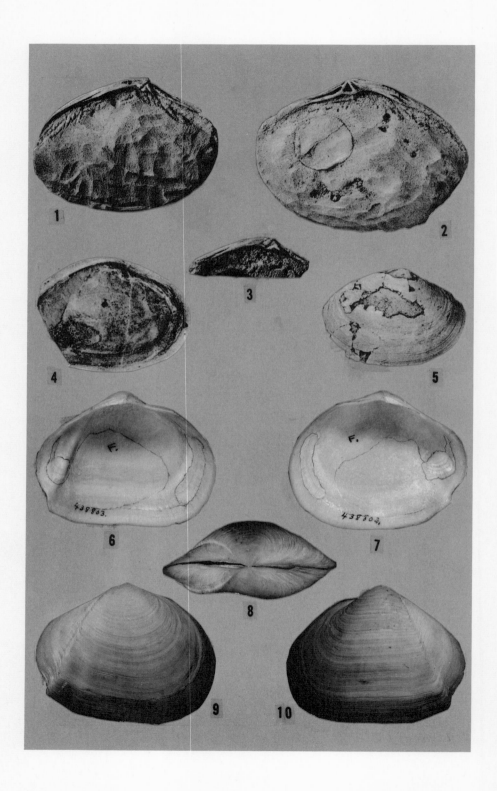

PLATE 42.

All figures × 1, except figures 1–3 (× 1½)

Figures   Page

Nelltia stenzeli Stephenson 101
  (Copied from Stephenson, 1952).

1. Hinge of left valve of paratype (U.S.N.M. 105433).
2. Interior of holotype, right valve (U.S.N.M. 105431).
3. Hinge of left valve of paratype (U.S.N.M. 105435a).
4. Incomplete right valve of paratype (U.S.N.M. 105435b).
5. Exterior of holotype (U.S.N.M. 105431).

Psammothalia cognata (C. B. Adams). 107
  Florida: U.S.N.M. 438803

6. Interior of left valve.
7. Interior of right valve.
8. Dorsal view.
9. Exterior of right valve.
10. Exterior of left valve.

PLATE 43.

Figures 1–4, × 3; figure 5, 4½; figure 6, × 1½; figures 7–9, × 1

Figures   Page

Quadrans (Acorylus) suberis (Dall). 99
   Pliocene, Shell Creek, Charlotte Co., Florida; U.S.G.S. 3300, U.S.N.M. 163446.

1–2. Exterior and interior of left valve.
3–4. Exterior and interior of right valve.

Strigilla (Simplistrigilla) strata Olsson. 102
   (Copied from Olsson, 1961).

   5. Exterior of holotype, a left valve.

Bathytellina citrocarnea (Habe)   99
   (Copied from Habe, 1964).

   6. Exterior of right valve.

Bendemacoma nigeriensis (Newton) 105
   (Copied from Newton, 1922).

   7. Exterior of lectotype, a left valve.
   8. Interior of right valve, a paralectotype.
   9. Interior of left valve, a paralectotype.

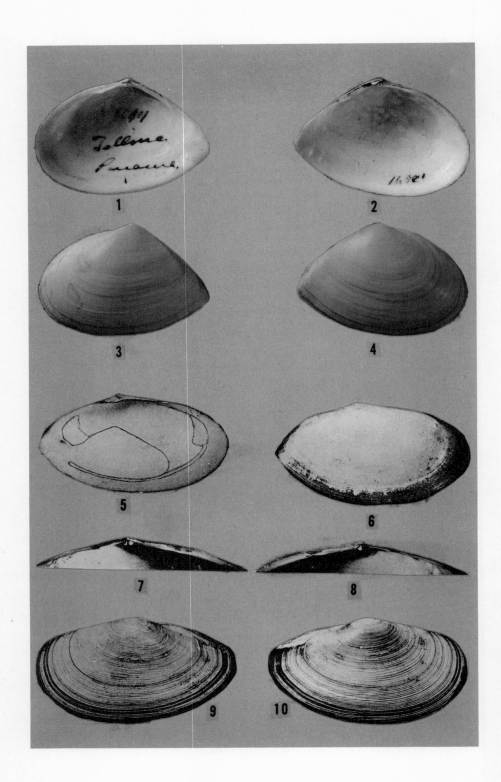

PLATE 44.

Figures 1–4, × 1; figures 5–6, × ⅚; figures 7–10, × 1½

Figures   Page

Tellina (Elpidollina) decumbens Carpenter. 98
  Panama, syntype (U.S.N.M. 16101)

1–4. Interior and exterior views.

Macoma (Ardeamya) columbiensis (Hanley) 104
  (Copied from Olsson, 1961).

  5. Interior of left valve.
  6. Exterior of right valve.

Tellina (Hertellina) nicoyana Hertlein and Strong 97
  (Copied from Hertlein and Strong, 1949).

  7. Hinge of right valve of holotype.
  8. Hinge of left valve of holotype.
  9. Exterior of left valve of holotype.
  10. Exterior of right valve of holotype.

PLATE 45.

All figures × 1

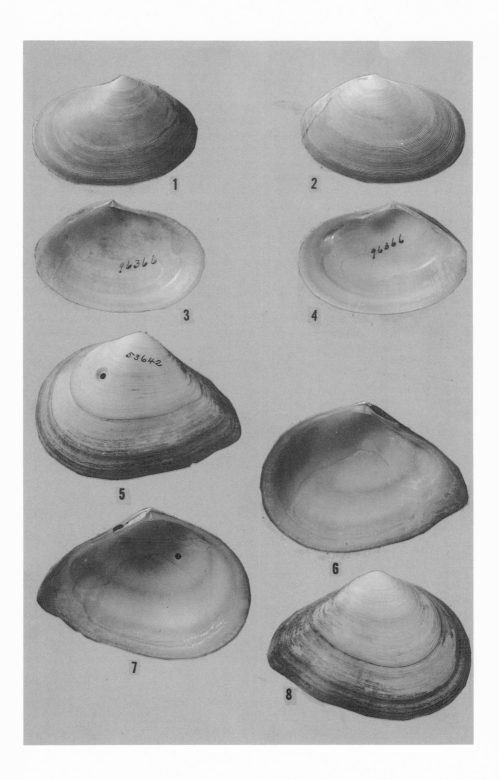

Bibliography

Adams, C. B., 1852.

Adams, H. and A., 1854–58, The genera of recent Mollusca: v. 2, London, 661 p.

Adams, Henry, 1860, On two new genera of acephalous mollusks: Proc. Zool. Soc. London, pt. 28, p. 369.

—— 1870, Descriptions of twenty-six new species of shells collected by Robert Mac'Andrew, Esq., in the Red Sea: Proc. Zool. Soc. London 1870, p. 788–793, pl. 48.

Adanson, Michel, 1757, Histoire naturelle du Sénégal: Coquillages, Paris, 275 p., 19 pl.

Agassiz, J. L. R., 1846, Nomenclator Zoologicus, ——: Fasc. IX & X. Mollusca, 98 p.

—— 1848, Nomenclatoris Zoologici Index Universalis ——: Solothurn, 1135 p.

Anton, Hermann E., 1837, Diagnosen einiger neuen Conchylien-Arten: Wiegmann's Arch. Naturges., v. 3, p. 281–286.

Bartrum, J. A., 1919, New fossil Mollusca: Trans. New Zealand Inst., v. 51, p. 96–100, pl. 7.

Bayan, J. F. F., 1870–73, Études faites dans la collection de l'Ecole des Mines sur des fossiles nouveaux ou mal connus: 2 Fasc. Paris, 162 p., 20 pl.

Bertin, Victor, 1878, Révision des Tellinidés du Museum d'Histoire Naturelle: Nouv. Arch. Mus. Nat. Hist. Nat. Paris, ser. 2, v. 1, p. 201–361, pl. 8–9.

Blainville, 1825, Manu. Malac. Vol. 1.

Bory de St. Vincent, J. B. G. M., 1827, Encyclopedie Methodique. Tableau encyclopedique et Methodique des trois regnes de la nature: Vers, Coquilles, Mollusques, et Polypiers. p. 83–84 (reprint), 133–180.

Bosc, L. A. G., 1801, Histoire naturelle des coquilles: Paris, 5 vols.

Boss, Kenneth J., and Kenk, Vida C., 1964, Anatomy and relationships of *Temnoconcha brasiliana* Dall: Occ. Pap. Moll. Dept. Moll., Harvard Univ., v. 2, n. 30, p. 325–343, pl. 56–60.

Brocchi, G., 1814, Conchiologia fossile subapennina: Milan, 2 vols.

Broderip, W. J., and Sowerby, G. B. 1829. Observations on new or interesting Mollusca contained, for the most part, in the Museum of the Zoological Society: Zool. Journal, v. 4, p. 363, fig. 15.

Broderip, W. J., and Sowerby, G. B., 1839, Voy. Beechy., Zool. p. 153, pl. 42, fig. 2.

Bronn, Heinrich G., 1848–49, Handbuch einer Geschichte der Natur., v. 3: Index Paleontologicus, Stuttgart, 3 vols.

Brown, Thomas, 1827, Illustrations of the Conchology of Great Britain and Ireland: London, V p., 52 pl.

Bruguiere, J. G., 1791–97, Encyclopedie Methodique: Tableau Encycl. et Method. des Trois Regnes de la Nature. Vers, Coquilles, Mollusques et Polypoeri. Paris, 132 p., 286 pl.

Bryan, William A., 1915, Natural History of Hawaii: Honolulu, 596 p., 117 pl.

Carpenter, Philip P., 1856, Description of new species of shells collected by Mr. T. Bridges in the Bay of Panama and its vicinity, in the collection of Hugh Cuming, Esq.: Proc. Zool. Soc. London, pt. 24, p. 159–171.

—— 1863, A supplementary report on the present state of our knowledge with regard to the Mollusca of the west coast of North America: Rep. Brit. Assoc. Advance Sci. 1863, p. 517–686.

Chemnitz, J. H., 1782, Neues systematisches Conchylien-Cabinet, vol. 6, 375 p., 36 pl.

—— 1795, Neues systematisches Conchylien-Cabinet, vol. 11, 310 p., pl. 174–213.

Children, J. G., 1823, Lamarck's genera of shells, translated from the French: London, 177 p.

Conrad, T. A., 1837, Descriptions of new marine shells, from Upper California: Collected by Thomas Nuttall, Esq., Journ. Acad. Nat. Sci. Phila., v. 7, p. 227–268, pl. 17–20.

—— 1860, Descriptions of new species of Cretaceous and Eocene Fossils of Mississippi and Alabama: Journ. Acad. Nat. Sci. Phila., ser. 2, v. 4, p. 275–298, pl. 46–47.

—— 1870, Notes on recent and fossil shells, with descriptions of new species: Am. Journ. Conch., v. 6, p. 71–78.

—— 1873, Descriptions of new genera and new species of fossil shells of North Carolina, in the State Cabinet at Raleigh: *in* Kerr Rep. Geol. Survey North Carolina, v. 1, App. A, p. 1–28. (As separate, 1873; *in* v., 1875).

Cossmann, M., 1885, Catalogue illustré des coquilles fossiles de l'Éocène des environs de Paris: Fasc. 1. Ann. Soc. Roy. Malac. Belgique, v. 21, p. 17–186, pl. 1–8.

—— 1892, Catalogue illustré des coquilles fossiles de l'Éocène des environs de Paris: Fasc. 5. Ann. Soc. Royale Malac. Belgique, v. 26, p. 3–163, pl. 1–3. (Separate).

—— 1902, Rectifications de Nomenclature: Rev. Crit. Paleozoologie, v. 6, p. 52.

—— 1921, Synopsis illustré des mollusques de l'Éocène et de l'Oligocène en Aquitaine: Mem. Soc. Geol. France. Paleont. Mem. No. 55, 220 p., 15 pl.

Couthouy, Joseph P., 1838, Descriptions of new species of Mollusca and shells, and remarks on several polypi found in Massachusetts Bay: Journ. Boston Soc. Nat. Hist., v. 2, p. 53–111, pl. 1–3.

Crosse, H. and Debeaux, O. Diagnoses d'éspèces nouvelles du nord de la Chine: Journ. de Conch., v. 11, p. 77–79.

Da Costa, E. M., 1779, Historia Naturalis Testaceorum Britanniae, or, The British Conchology ——: London, 254 p., 17 pl.

Dall, W. H., 1889, A Preliminary catalogue of the shell-bearing marine mollusks and brachiopods of the south-eastern coast of the United States, ——: Bull. U.S. Nat. Mus. No. 37, 221 p., 74 pl.

—— 1900a, Synopsis of the family Tellinidae and of the North American species: Proc. U.S. Nat. Mus., v. 23, p. 285–326, pl. 2–4.

—— 1900b, Contributions to the Tertiary Fauna of Florida ——: Part V. Trans. Wagner Free Inst. Sci. Phila., v. 3, pt. 5, p. 949–1218, pl. 36–47.

—— 1921, Two new South American shells: Nautilus, v. 34, p. 132–133.

—— 1924, Notes on molluscan nomenclature: Proc. Biol. Soc. Washington, v. 37, p. 87–90.

Dall, W. H., Bartsch, Paul, and Rehder, H. A., 1939, A manual of the recent and fossil marine pelecypod mollusks of the Hawaiian Islands: Bernice P. Bishop Mus. Bull. 153, 233 p., 58 pl.

Dance, S. P. and Eames, F. E., 1966, New molluscs from the recent Hammar Formation of south-east Iraq: Proc. Malac. Soc. London, v. 37, pt. 1, p. 35–43, pl. 2–5.

d'Argenville, A. J. Dezallier, 1742, L'Histoire Naturelle eclaircie dans deux de ces parties principales, la Lithologie et la Conchyliologie, ——: Paris, 492 p., 33 pl. + 14 p., 3 pl.

Dautzenberg, P. and Fischer, H., 1907, Contribution a la faune malacologique de l'Indo-chine: Journ. Conch., v. 54, p. 145–226, pl. 5–7.

Delessert, Benjamin, 1841, Recueil de Coquilles décrites par Lamarck —— et non encore figurées: Paris, 40 pl. and explanations.

Deshayes, G. P., 1824, Description des coquilles fossiles de environs de Paris: v. 1. Conchiferes. Paris, 392 p., 65 pl.

—— 1832, Encyclopedie Methodique: Histoire Naturelle des Vers, v. 3, p. 595–1152.

—— 1835–45, Lamarck, Histoire Naturelle des Animaux sans Vertebres: Deuxieme Ed., vols. 6–11, Histoire de Mollusques, Paris.

—— 1845–48, Exploration scientifique de l'Algerie —— 1840-42, Zoologie: Histoire Naturelle des Mollusques. XX + 609 p.; Atlas 160 p., 155 pl.

—— 1854, Descriptions of new shells from the collection of Hugh Cuming, Esq: Proc. Zool. Soc. London, pt. 22, 1854, p. 317–371.

—— 1860, Description des animaux sans vertebres decouverts dans le Bassin de Paris: v. 1. Mollusques Acephales Dimyaires. Paris, 912 p.; Atlas 88 p., 89 pl.

Donovan, E., 1800–04, The Natural History of British Shells, ——: London, 5 vols.

d'Orbigny, Alcide, 1849–52, Prodrome de Paleontologie stratigraphique universelle des Animaux Mollusques et Rayonnes, ——: Paris, 3 vols.

Eames, F. E., 1957, Eocene Mollusca from Nigeria: A revision: Brit. Mus. (Nat. Hist.) Bull., Geology, v. 3, no. 2, p. 25–70, pl. 5–6.

Fabricius, Otto, 1780, Fauna Groenlandica, ——: Copenhagen, 452 p., 1 pl.

Finlay, H. J., 1926, A further commentary on New Zealand Molluscan systematics: Trans. N.Z. Inst., v. 57, p. 320–485, pl. 18–23.

Fischer, Paul, 1880–87, Manual de Conchyliologie et de Paleontologie conchyliologique, ——: Paris, 1368 p., 23 pl.

Fleming, John, 1828, History of British Animals: Edinburgh, 565 p.

Gardner, Julia, 1916, Systematic Paleontology. Upper Cretaceous. Mollusca: Maryland Geol. Survey Bull., Upper Cretaceous, p. 371–732, pl. 12–45.

—— 1928, The molluscan fauna of the Alum Bluff Group of Florida. Part V. Tellinacea, etc.: U.S. Geol. Survey Prof. Paper 142-E, p. 185–249, pl. 29–36.

Gatliff, J. H. and Gabriel, C. J., 1914, Alterations in the nomenclature of some Victorian Marine Mollusca: Victorian Naturalist, v. 31, p. 82–84.

Gistel, 1848

Gmelin, J. F., 1791, Systema Naturae: Ed. 13, v. 1, pt. 6, p. 3021–3909.

Goldfuss, G. A., 1826–44, Petrefacta Germaniae, ——Abbildungen und Beschreibungen der Petrefacten Deutschlands: Düsseldorf, 3 vols.

Gould, Augustus A., 1841, Report on the Invertebrata of Massachusetts: Cambridge. 373 p., 15 pl.

Gray, J. E., 1825, A list and description of some species of shells not taken notice of by Lamarck: Ann. Phil., v. 25, p. 135–140, 407–415.

—— 1842, Synopsis of the contents of the British Museum: London, 44th ed.

—— 1847, A list of the genera of Recent Mollusca, their synonyma and types: Proc. Zool. Soc. London, pt. 15, p. 129–219.

—— 1851, List of the specimens of British Animals in the collection of the British Museum, Part VII: Mollusca Acephala and Brachiopoda, 167 pp.

Gualtieri, N., 1742, Index testarum conchyliorum quae adservantur in Museo N. Gualtieri ——: Florence, 23 p., 110 pl.

Habe, Tadashige, 1951–53, Genera of Japanese Shells: Pelecypoda, n. 1–4, 326 p.

—— 1958, Report on the Mollusca chiefly collected by the S. S. Soyo-Mara of the Imperial Fisheries Experimental Station on the continental shelf bordering Japan during the years 1922–30. Part 4, Lamellibranchis (2): Publ. Seto Marine Lab., v. 7, n. 1, p. 19–52, pl. 1–2.

—— 1961, Coloured Illustrations of the shells of Japan (II): Osaka. 183 p., 66 pl. (in Japanese)

—— 1964, Shells of the western Pacific in color: Osaka. v. II. 233 p., 60 pl.

Habe, T. and Ito, Kiyoshi, 1965, Shells of the World in Colour. v. I. The Northern Pacific: Osaka, viii [2 maps], 176 p., 56 pl.

Hanley, Sylvanus, 1844–45, Descriptions of new species of Tellina, collected by H. Cuming: Proc. Zool. Soc. London, pt. 12, p. 59–64, 68–72, 140–144, 146–149, 164–166.

—— 1847, A Monograph of the genus Tellina: G. B. Sowerby, Thesaurus Conch., v. 1, p. 221–336, pl. 56–66.

Harris, George F. and Burrows, Henry W., 1891, The Eocene and Oligocene beds of the Paris Basin: Geologist's Association. London, 129 p.

Hedley, C., 1913, Studies on Australian Mollusca: Part XI, Proc. Linn. Soc. New South Wales, v. 38, p. 258–339, pl. 16–19.

Hermannsen, A. N., 1846–49, Indicis-Generum Malacozoorum Primordia: Cassel, 2 vols.

Hertlein, L. G., and Strong, A. M., 1949, Eastern Pacific Expeditions of the New York Zoological Society. XL. Mollusks from the West Coast of Mexico and Central America. Part VII: Zoologica, v. 34, pt. 2, p. 63–97, pl. 1.

Iredale, Tom, 1915, A commentary of Suter's "Manual of the New Zealand Mollusca": Trans. New Zealand Inst., v. 47, p. 417–497.

—— 1927, New molluscs from Vanikoro: Rec. Australian Mus., v. 16, p. 73–78, pl. 5.

—— 1929, Queensland molluscan notes, no. 1: Mem. Queensland Mus., v. 9, p. 261–297, pl. 30–31.

—— 1930, Queensland molluscan notes, no. 2: Mem. Queensland Mus., v. 10, p. 73–88, pl. 9.

—— 1930, More notes on the marine mollusca of New South Wales: Rec. Australian Mus., v. 17, p. 384–407, pl. 62–65.

—— 1936, Australian shell notes, no. 2: Rec. Australian Mus., v. 19, p. 267–340, pl. 20–24.

—— 1937, The Middleton and Elizabeth Reefs, South Pacific Ocean: Mollusca. Australian Zool., v. 8, p. 232–261, pl. 15–17.

Klein, Jacob T., 1753, Tentamen Methodi Ostracologie, ——: Leiden, 282 p., 12 pl.

Lamarck, J. B. P. A., de M. de, 1798–1816, Encyclopedie Methodique: Tableau encyclopedique et methodique des trois regnes de la nature. Vers, Coquilles, Mollusques, et Polypiers. Liste des objets 16 p., pl. 287–488.

—— 1799, Prodrome d'une nouvelle classification des coquilles: Mem. Soc. Hist. Nat. Paris, v. 1, p. 63–91.

—— 1801, Système des animaux sans vertèbres ——: Paris, 432 p.

—— 1802–06, Sur des fossiles des environs de Paris, comprenant la détermination des espèces qui appartiennent aux animaux marins sans vertèbres, et dont la plupart sort figurées dans la collection des velins des Muséum: Ann. Mus. Hist. Nat. Paris, v. 1, p. 299–312, 383–391, 474–479; v. 2, p. 57–64, 163–169, 217–227,

315–321, 385–391; v. 3, p. 163–170, 266–274, 343–352, 436–441; v. 4, p. 46–55, 105–115, 212–222, 289–298, 429–436; v. 5, p. 28–36, 91–98, 179–180, 237–245, 349–356; v. 6, p. 117–126, 214–221, 222–228, 337–345, 407–415; v. 7, p. 53–62, 130–140, 231–242, 419–430; v. 8, p. 156–166, 347–355, 461–469.

—— 1805–09, Sur les Fossiles des environs de Paris. Explication des planches relatives aux coquilles fossiles des environs de Paris. Ann. Mus. Hist. Nat. Paris, v. 6, p. 222–228, pl. 1–4; v. 7, p. 242–244, pl. 5–7; v. 8, p. 77–79, 383–388, pl. 8–14; v. 9, p. 236–240, 399–401, pl. 15–20; v. 12, p. 456–459, pl. 21–24; v. 14, p. 374–375, pl. 25–28.

—— 1818, Histoire naturelle des animaux sans vertèbres, v. 5: Paris, 612 p.

Lamy, Edward, 1918, Les Tellines de la Mer Rouge (d'après les matériaux recueillis par M. le Dr. Jousseaume.) Bull. Mus. Hist. Nat. Paris, v. 24, p. 26–33, 116–123, 167–172.

Laws, C. R., 1933, New tertiary mollusca from Turiaru District, South Canterbury, New Zealand: Trans. New Zealand Inst., v. 63, p. 315–329, pl. 29–33.

Leach, W. E., 1819, A list of invertebrate animals, discovered by His Majesty's Ship Isabella, in a voyage to the Arctic Regions: In John Ross, A voyage of discovery, ——. App. II., Zoological Memoranda, p. LXI–LXIV.

—— 1852, A synopsis of the Mollusca of Great Britain: London, 376 p., 13 pl.

Linnaeus, C., 1758, Systema Naturae per regna tria naturae, ——: v. 1, Stockholm 824 p.

Lister, Martin, 1678, Historiae Animalium Angliae tres tractatus. London, 250 p., 9 pl.

Mac'Andrew, Robert, 1857, Report on the marine testaceous mollusca of the northeast Atlantic and neighbouring seas, and the physical conditions affecting their development: Rep. Brit. Assoc. Advancement Sci. 1863, p. 99–158.

Martyn, Thomas, 1784–87, The Universal Conchologist, ——: London, 4 vols. 160 plates.

Marwick, J., 1924, Zealeda and Barytellina, new fossil molluscan genera from New Zealand: Proc. Mal. Soc. London, v. 16, p. 25–26.

—— 1928, The Tertiary Mollusca of the Chatham Islands including a generic revision of the New Zealand Pectinidae: Trans. New Zealand Inst., v. 58, p. 432–506.

—— 1929, Tertiary Molluscan fauna of Chatton, Southland: Trans. New Zealand Inst., v. 59, p. 903–934, 75 figs.

—— 1934, Some new New Zealand Tertiary Mollusca: Proc. Malac. Soc. London, v. 21, p. 10–21, pl. 1–2.

Meek, F. B., 1864, Check list of the invertebrate fossils of North America. Cretaceous and Jurassic: Smithsonian Misc. Coll. No. 177. 40 p.

—— 1871, Preliminary paleontological report, consisting of lists of fossils, with description of some new types, etcetera: U.S. Geol. Survey Wyoming, Prelim. Rep., (4th Ann. Rep. Wyoming and Territories) p. 287–318.

Melvill, J. C., 1893, Descriptions of twenty-five new species of marine shells from Bombay: Collected by Alexander Abercrombie, Esq., Mem. Proc. Manchester Lit. Phil. Soc., ser. 4, v. 7, p. 52–67, 1 pl.

Menke, K. T. and Pfeiffer, L., 1861, Eronterte Mollusken [Index], Malak. Blätt, v. 7, p. iv–viii.

—— 1861, Beiträge zur Molluskenfauna Central-Amerika's (Fortsetzung): Malak. Blätt., v. 7, p. 170–213.

Montagu, George, 1803, Testacea Britannica or natural history of British shells,——: London, 606 p., 10 pl.

Monterosato, A. de, 1884, Nomenclatura generica e specifica di alcune conchiglie Mediterranee: Palermo, 152 p.

Mörch, O. A. L., 1852, Catalogus Conchyliorum—Comes de Yoldi,——: Fasc. 2. Acephala., Copenhagen, 74 p.

—— 1860, Malakozool. Blätt, Vol. 7.

Mühlfeld, J. K. Megerle von, 1811, Entwurf eines nueen system's der Schalthiergehause: Ges. Naturf. Fr. Berlin, Mag. neuesten Entdeck. gesamm Naturk., v. 4, p. 38–72.

Newton, R. Bullen, 1922, Eocene Mollusca from Nigeria: Geological Survey of Nigeria, Bull., no. 3, 114 p., 11 pl.

Olsson, A. A., 1942, Tertiary and Quaternary fossils from the Burica Peninsula of Panama and Costa Rica: Bull. Amer. Paleont., v. 27, p. 157–258, pl. 14–25.

—— 1944, Contributions to the paleontology of Northern Peru, Part VII, The Cretaceous of the Paita Region: Bull. Amer. Paleont., v. 28, 146 p., 17 pl.

—— 1961, Mollusks of the tropical eastern Pacific, particularly from the southern half of the Panamic-Pacific Faunal Province (Panama to Peru), Panamic-Pacific Pelecypoda: Paleont. Res. Inst., Ithaca, 574 p., 86 pl.

Olsson, A. A., and Harbison, Anne, 1953, Pliocene Mollusca of southern Florida, with special reference to those of North Saint Petersburg: Acad. Nat. Sci. Phila. Mon., no. 8, 457 p., 65 pl.

Pallary, G., 1920, Malacologie: Exploration scientifique du Maroc, Fasc. 2. 107 p., 1 pl.

Payraudeau, B. C., 1827, Catalogue descriptif et methodique des Annelides et des Mollusques de l'Ile de Corse: Paris. 218 p., 8 pl.

Pennant, Thomas, 1776–77, British Zoology, Ed. 4: Warrington and London, 4 vols.

Philippi, R. A., 1842–51, Abbildungen und Beschreibungen neuer oder wenig bekannter Conchylien, ——: Cassel, 3 vols.

—— 1846, Diagnosen einiger neuen Conchylienarten: Zeitschr. Malakozool., v. 3, p. 19–24, 49–55.

Pilsbry, H. A., 1918, Marine Mollusks of Hawaii, IV–VII: Proc. Acad. Nat. Sci. Phila., v. 69, p. 309–333, pl. 20–22.

Pilsbry, H. A. and Lowe, H. N., 1932, West Mexican and Central American mollusks collected by H. N. Lowe, 1929–31: Proc. Acad. Nat. Sci. Phila., v. 84, p. 33–144, pl. 1–17.

Pilsbry, H. A. and Olsson, A. A., 1941, A Pliocene fauna from western Ecuador: Proc. Acad. Nat. Sci. Phila., v. 93, p. 1–79, pl. 1–19.

Poli, G. S., 1791, Testacea utriusque Siciliae eorumque historia et anatome: v. 1, 214 p. Parma.

Pulteney, R., 1799, Catalogues of the birds, shells, and some of the more rare plants of Dorsetshire: Hutchin's History of that country. London, 92 p.

Quoy, J. R. C. and Gaimard, J. P., 1832–35, Voyage—de l'Astrolabe—: Zoologie, vols. 2–3 — Atlas: Mollusques. Poissons.

Reeve, Lovell A., 1866–69, Tellina: Conchologica Iconica, v. 17, 58 pl.

Renieri, S. A., 1804, Tavolo alfabetica delle Conchiglie Adriatiche, ——: Padua, p. V–XIII.

Röding, P. F., 1798, Museum Boltenianum——: Hamburg, 199 p.

Römer, E., 1872, Die Familie der Tellmuscheln, Tellinidae: Syst. Conch-Cat., ed. 2, v. 10, pt. 4, p. 1–291, pl. 1–52.

Rumphius, G. E., 1711, Thesaurus imaginum Piscium Testaceorum; etc.——: Leiden, 23 p., 60 pl.

—— 1741, D'Amboinsche Rariteitkamer, ——: Amsterdam, 383 p., 60 pl.

Sacco, Federico, 1901, I Molluschi dei terreni terziarii del Piemonti e della Liguria. Parte 29, Turin, 216 p., 29 pl.

Salisbury, A. E., 1929, A twice pre-occupied name: Proc. Mal. Soc. London, v. 18, p. 255.

—— 1934, On the nomenclature of Tellinidae, with description of new species and some remarks on distribution. Proceedings of the Malacological Society of London, Vol. 21, Part 2.

—— 1929, Polymetis: Proc. Mal. Soc. London, v. 18, p. 258.

—— 1934, On the nomenclature of Tellinidae, with descriptions of new species and some remarks on distribution: Proc. Mal. Soc. London, v. 21, p. 74–91, pl. 9–14.

Say, Thomas, 1827, Descriptions of Marine shells recently discovered on the coast of the United States: Journ. Acad. Nat. Sci., Philadelphia, v. 5, p. 207–221.

—— 1830–34. American Conchology, ——: New Harmony, 60 pl., with text.

Schmidt, F. C., 1818, Versuch über die beste Einrichtung——, vorzüglich der Conchylien-Sammlungen, ——: Gotha, 252 p.

Schlotheim, 1820

Schrenck, Leopold von, 1861, Vorläufige Diagnosen einiger neuer Molluskenarten aus der Meerenge der Tartarei und dem Nordjapanischen Meere: Bull. Acad. Imp. Sci., St. Petersbourg, v. 4, p. 408–413.

—— 1867, Mollusken des Amur-Landes und des Nordjapanischen Meeres: Reisen u. Forsch. im Amur-Lande 1854–56——., v. 2, Zool., p. 259–974, pl. 12–30.

Schumacher, C. F., 1817, Essai d'un nouveau système des habitations des vers testacés: Copenhagen, 277 p., 22 pl.

Scopoli, G. A., 1777, Introductio ad historiam naturale, ——: Prague, 506 p.

Smith, E. A., 1885, Report on the Lamellibranchiata: Rep. Sci. Res. Voyage H.M.S. *Challenger*, Zool. v. 13, pt. 35, 341 p., 25 pl.

—— 1891. On a collection of marine shells from Aden, with some remarks upon the relationship of the Molluscan fauna of the Red Sea and the Mediterranean: Proc. Zool. Soc. London, 1891, p. 390–436, pl. 33.

Sowerby, G. B., I, 1825, A catalogue of the shells contained in the collection of the late Earl of Tankerville, ——: London, p. 92, 9 pl.

—— 1839, Molluscous Animals, and their shells: Zoology of Captain Beechey's voyage; ——, p. 101–155, pl. 33–44.

Sowerby, G. B., II, 1873. Monograph of the genus Panopaea: Reeve, Conch. Icon., v. 19, Panopaea, 6 pl.

Sowerby, G. B., III, 1909, Descriptions of new species of *Terebra, Pleurotoma, Trochus, Tellina, Dosinia,* and *Modiola:* Proc. Mal. Soc. London, v. 8, p. 198–201.

Sowerby, James, 1804–06, The British miscellany: or coloured figures of new, rare, or little-known animal subjects; ——: London, 2 vols.

Spengler, Lorenz, 1798, Over det toskallede slaegt Tellinerne: Skriv. Naturhist. Selskabet (Kiobenhavn), v. 4, pt. 2, p. 67–121, pl. 12.

Stearns, Robert E. C., 1873, Shells collected at San Juanico, lower California, by William M. Gabb: Proc. California Acad. Sci., v. 5, p. 131–132.

—— 1894, The shells of the Tres Marias and other localities along the shores of lower California and the Gulf of California: Proc. U.S. Nat. Mus., v. 17, p. 139–204.

Stephenson, Lloyd W., 1923, Invertebrate fossils of the Upper Cretaceous formations: North Carolina Geol. Econ. Survey., v. 5, pt. 1, p. 1–402, pl. 1–102.

—— 1952, Larger invertebrate fossils of the Woodbine Formulation (Cenomanian) of Texas: Geol. Survey Prof. Paper 242, 226 p., 59 pl.

Stoliczka, Ferdinand, 1870–71, Cretaceous fauna of southern India: The Pelecypoda, v. 3, Palaeontologica Indica. Mem. Geol. Survey of India. Calcutta, 537 p., 50 pl.

Thiele, Johannes, 1934, Handbuch der systematischen Weichtierkunde: Pt. 3, p. 779–1022.

Thiele, Johannes, and Jäeckel, Siegfried, 1931, Muscheln der Deutschen Tiefsee-Expedition: Wissensch. Ergebn. Deutsch. Tiefsee-Exped., v. 21, pt. 2, p. 159–268, pl. 6–10.

Thorpe, Charles, 1844, British Marine Conchology, ——; London, 267 p., 8 pl.

Tryon, George W., Jr., 1869, Catalogue of the family Tellinidae: Amer. Journ. Conch., v. 4, p. 72–126.

—— 1874, American Marine Conchology: ——, Philadelphia, 208 p., 44 pl.

—— 1882–84, Structural and Systematic Conchology: Philadelphia, 3 vols.

Turton, William, 1819, A conchological Dictionary of the British Islands: London, 272 p., 28 pl.

—— 1822, Conchylia Insularum Britannicarum. The shells of the British Islands, systematically arranged: Dithyra. London, 279 p., 20 pl.

von Born, Ignaz, 1778, Index rerum naturalium musei caesarei vindobonensis, pars I: Testacea. Vienna, [38] + 458 + [82] p., 1 pl.

—— 1780. Testacea musei caesarei vindobonensis: Vienna, 442 p., 18 pl.

Weinkauff, H. C., 1867–68, Die Conchylien des Mittelmeeres, ihre geographische und geologische Verbreitung: Cassel, 2 vols.

Whitfield, R. P., 1885, Brachiopoda and Lamellibranchiata of the Raritan days and greens and marls of New Jersey: U.S. Geol. Survey Mon. 9, 269 p., pl. 1–28.

Winkworth, R., 1932, The British Marine Mollusca: Journ. of Conch., v. 19, p. 211–252.

Wood, W., 1815, General Conchology; ——: London, 246 p., 59 pls.

—— 1818. Index Testaceologicus or A Catalogue of Shells, ——: London, 188 p., 8 pl.

Yamamoto, Gotarô, and Habe, Tadashige, 1959, Fauna of shell-bearing mollusks in Mutsu Bay Lamellibranchia (2): Bull. Mar. Biol. Sta. Asamushi, Tôhoku Univ., v. 9, no. 3, p. 85–122, pl. 6–14.

Yonge, C. M., 1949, The sea shore: London, 311 p.

Author Index

Subject Index

211